Community, Food Insecurity, and a Global Perspective on Campus Food Pantries

"As someone who has had the privilege of supporting Dr. Sonya Sharififard throughout her research, I can say with confidence that this book is the result of her passionate commitment to shedding light on the challenges that students face in accessing nutritious food. An insightful and thoroughly researched examination of food insecurity on college campuses, this book is essential reading for educators, policymakers, and anyone who is dedicated to fostering equitable educational environments and inspiring action towards meaningful and informed solutions."

—Farzin Madjidi, EdD Dean, *Graduate School of Education and Psychology, Professor of Leadership Pepperdine University*

"Dr. Sharififard gives us a multi-dimensional systematic look at the corrosive and long-lasting impacts of food insecurity among students in higher education and unequal ways those impacts are felt. Her findings should be of interest to anyone concerned about making our higher education institutions (and society as a whole!) more inclusive, just, and vibrant."

—Amy Guptill, PhD Professor, *SUNY Brockport*

"I wholeheartedly endorse Dr. Sonya Sharififard's groundbreaking book on food insecurity on college campuses. This essential work sheds light on the critical issue of food accessibility and its profound impact on the lives of students. Through a multidisciplinary lens, Dr. Sharififard masterfully weaves together insights from food studies, educational leadership, and human geography, presenting a rich tapestry of the cultural and social complexities surrounding food practices. Each chapter highlights pressing issues and offers actionable resources and activities that empower readers to engage with these challenges meaningfully. In an era where students are often forced to choose between academic necessities and basic sustenance, this book invites us to consider the intersection of identity, belonging, and public health, ultimately advocating for a more equitable food system. I strongly recommend this book to educators, policymakers, campus communities and anyone passionate about creating a more just and nourishing environment for our students."

—Dr. Tadios Belay, President and CEO *U.S. Africa Institute*

"Sharififard's book elevates cultural dimensions of efforts to build nourishing institutional food environments. What curricular and procurement practices meet needs of groups increasingly targeted to fill declining classes, as students arrive with increasing needs for culturally specific diets, and dietary needs not helped by increasingly ultra processed and chemically-contaminated foods?"

—Valentine Cadieux, PhD Associate Professor-Anthropology, *Hamline University, President of the Agriculture, Food, and Human Values Society*

Sonya Sharififard

Community, Food Insecurity, and a Global Perspective on Campus Food Pantries

palgrave
macmillan

Sonya Sharififard
Graduate School of Education and Psychology
Pepperdine University
Malibu, CA, USA

ISBN 978-3-031-78222-0 ISBN 978-3-031-78223-7 (eBook)
https://doi.org/10.1007/978-3-031-78223-7

Cover illustration: Pattern © Melisa Hasan

This Palgrave Macmillan imprint is published by the registered company Springer Nature Switzerland AG.
The registered company address is: Gewerbestrasse 11, 6330 Cham, Switzerland

Preface

The study and lived experiences of campus food pantries are often numerous, and significant to the understanding of environmental, socioeconomic, and sociological complexities that shape communities and develop geographically dispersed populations. The differing student and neighboring demographics across college and university campuses globally and the local food environments impact the extent to which student resources are used throughout campuses and by several different actors. Additionally, the cultural and ecological differences of each institution may not be fully indicative of diverse food cultures, characteristics, and lifestyles among a specific community, jurisdiction, or region. The broad categories of food items would also make it challenging to infer meaningful connections about the overall quality consistent in a food pantries operations; but are used to identify demographic and environmental changes in education, policy, and public health.

A food pantry is a distribution center in which food, beverages, and utensils are kept and supplied for individuals in need. They are also used as centers and spaces where supplies, toiletries, and accessories are available for personal or professional use. Food pantries have had significant and historic implications beginning with the world's first food bank at St. Mary's Food Bank Alliance, in 1967. Started in Phoenix, Arizona, the food bank system began as an entrepreneurial venture to prevent food waste and spoilage from traditional grocery operations. The idea quickly launched and helped shape decades of food aid and services around the world. Continuing in operation, it is now a member of the Feeding

America network, a nonprofit organization that serves over 200 local food banks. While food pantries became recognized on colleges campuses from the basis of the food banking networks, other types of food aid on college campuses have started from other types of support such as meal delivery, garages, homes where extracurricular activities and professional societies meet as a gathering place, and kitchen cabinets where patrons can also make meals and take what they need back home.

Often, campus food pantries have limited geographic reach, making it difficult to serve remote or hard-to-reach areas. In that same respect, food pantries may not always have the resources to distribute food equally, leading to disparities in access to food among different communities or populations. For example, the pantry may only receive a limited variety of food items, which can lead to a lack of nutritious options. Further, expiring food dates, damaged or spoiled products, or lack of storage space, the problem of food waste and environmental pollution persists over time.

With respect to eligibility requirements, students can often become excluded from receiving adequate services professionally. Staff and support volunteers may not be equipped to serve clients who speak different languages or may not have the necessary expertise or resources to provide balanced and nutritious food options which can negatively impact the health and wellbeing of students, also perpetuating food insecurity. Additionally, campus food pantries may have limited hours of operation or physical locations that are not easily accessible to all clients. Pantries may not be equipped to accommodate special dietary needs, such as gluten-free, kosher, or halal options, which can exclude certain individuals from accessing services. Pantries often rely heavily on volunteers, which can create inconsistent service levels and limit the ability to expand operations. Pantries may not have strong partnerships with other organizations or stakeholders, which can limit their ability to leverage resources and expertise. Pantries may lack adequate infrastructure, such as storage facilities, transportation options, or cooking facilities, which can hinder their ability to provide services. These limitations highlight the need for pantries to continually assess their operations and adapt to the evolving needs of their communities to provide effective and sustainable food assistance.

While quantitative research may prevent researchers from fully understanding the reasons and consequences of an occurrence, it is useful to determine the extent of the problem. The quantitative approach will also assist with the exploration of this study about the prevalence of campus food pantries. The effectiveness of on-campus food pantries could assess the extent to which higher education institutions advance student success and engagement. While it is difficult to understand the rate of food pantry usage across every institute of higher education around the world, some insights can be used to assess how such campus food pantries meet the dietary and nutritional needs of student populations. They can also be used to understand how they have improved student wellbeing and relations across rural partners, economic development, and performance improvement. When students feel comfortable with the knowledge, skills, and resources they have acquired earlier in life, they can create more opportunities for themselves and their communities.

Further, the complexity of the availability of cultural foods and other types of foods that a student might need for their diet would be virtually impossible to obtain. A student may want to visit a pantry to acquire products and items even if they are not experiencing food insecurity. They might also feel embarrassed about their food insecurity needs, so they will opt to choose food products that they believe would not result in judgment, criticism, or questioning from the pantry staff members or their peers, contributing to a deep sense of impostor syndrome where students feel self-doubt about their academic accomplishments and their ability to sustain a healthy and well-balanced lifestyle.

Finally, one cannot monitor food consumption when students leave the pantry. Multiple underlying assumptions of how individuals consume food on a social and personal matter might not adequately consider how contemporary urban life in most global cities is characterized by high degrees of mobility. Patrons might also have unique individual shopping preferences or dietary food preferences that are incompatible with the ideas and cultural influences that shape the practices and characteristics in their communities. Despite the above limitations, the most pertinent relationships between food insecurity and campus food pantries are explored, suggesting that the possibility of increased food stores in neighborhoods, such as convenience stores, can limit the types of foods available in the community and on campus. A variety of food stores are necessary to serve the diverse student populations enrolled at college campuses and universities. This book provides recommendations for the integration of advancing student success within campus food pantry system and the common good.

Malibu, CA, USA Sonya Sharififard

Acknowledgements

Thank you so much to Milana Vernikova, Ben Ingraham, Tikoji Rao, and the wonderful staff at Springer for their leadership, expertise, and support on this project. I am very grateful for their professionalism and belief in this process.

The author confirms contribution and has agreed to be accountable for all aspects of the work in ensuring that questions related to the accuracy or integrity of any part of the work are appropriately investigated and resolved.

Availability of Data and Materials

The data that support the findings of this study are available on request from the author.

Contents

CHAPTER 1

Introduction

Abstract This study explores whether foods available at campus food pantries can affect the cultural and social environment of the food retailing spaces around the campus and neighborhood community—an analysis of 75 college food pantries between 2019–2023 and a case study analysis. The lived experience of students impacted by the COVID-19 pandemic, participating in school, and meeting the needs of their undergraduate or graduate educational goals contributed to their higher learning outcomes when they experienced challenges with finding food that met their individual dietary needs on and off campus. While the food sources of many pantries were close to securing a food supply for immediacy, the long-term consequences of food insecurity continue to affect students' academic performance and sense of belonging in campus communities.

Keywords Food insecurity • Food pantry • SDGs • Kosher • Halal • Jain • Higher education leadership

The 2030 Agenda for Sustainable Development and its 17 Sustainable Development Goals (SDGs) was adopted by world leaders in 2015. (SDG) 2 is to "End hunger, achieve food security and improved nutrition and promote sustainable agriculture" by 2030. This goal aims to ensure that everyone has access to safe, nutritious food on a regular basis. Today,

S. Sharififard, *Community, Food Insecurity, and a Global Perspective on Campus Food Pantries*,
https://doi.org/10.1007/978-3-031-78223-7_1

almost 30% of the world's population is food insecure, with 45% unable to afford a nutritious, sustainable, and healthy diet. It has been estimated that around 600 million people will face hunger in 2030. This suggests that the Zero hunger SDG goal is far from being met.

An emerging area of concern is also among college and university students (Daugherty et al., 2019; Davis et al., 2020; Evans, 2016; Hagedorn & Olfert, 2018; McCarthy, 2019; McArthur et al., 2019; Mukigi & Brown, 2019; Ullevig et al., 2019) who face severe to chronic levels of food insecurity while completing their studies. Food insecurity refers to a condition in which an individual has insufficient or reduced intake, reduced or less quality, limited or no variety, or desirability in their diet, as well as disrupted eating patterns because of limited or no access to nutritious and safe foods. Generally, the most common forms of food insecurity challenges include either the inability to afford balanced and nutritional meals, the reduction in the size of meals, the conserving of available food, or the avoidance of eating regular meals because of a lack of funds or reliable income stream.

In 2019, a liberal arts student at a community college in a resort town in California visited the food pantry in the newly constructed Student Services building that was met with celebratory recognition and belonging a week prior. Across the picturesque campus and local skiing resorts, students gathered around the building, awaiting a ribbon cutting ceremony for the new campus food pantry. This unique kitchen space would equip students with meeting their basic needs, including toiletries, first aid products, personal care items such as shampoos, toothbrushes, and soap. The pantry was also designed to provide students with hope. The pantry was there to distribute food. It was a call to diversity of food insecurity, of challenges that meet students on a frequent basis on campus. In that same year, the world's largest educational research association gathered in Toronto. Attendees were stunned to find a food donation collection box in the main entrance at the lobby. With instructions pasted on the box, patrons could place food items to be distributed to local schools as well as university students facing food insecurity. On the cover of the box, several statistics were displayed showing the severity of the food insecurity problem around the world. On the other side of the box, a QR code was placed with instructions to become a member of the association's governing special interest division and participate in a focus group webinar where researchers and interested individuals could discuss the complexities of the problem.

A year later, the problem and issues surrounding food insecurity would be exacerbated by the global COVID-19 pandemic. Since 2001, Canada has had food pantries on four university campuses to help support students. Each pantry is concerned about the economic, ecological, and social implications of the food system. In 2023, 1.9 million visits were made to a food pantry. Most of the clients who used the services of the food banks were employed; but could not make ends meet. In 2021, The University of Pretoria in South Africa launched the Food Parcel Project in response to the significant increase in students who do not have access to healthy food and toiletries. The faculty of the health sciences division created a food parcel project to support students with food required to complete their education with the intent of developing a paying it forward culture among students and staff by encouraging those who can afford to donate non-perishable foods to support students from low-income households facing further financial strain due to the pandemic. The project also provided toiletries and personal care items that supported their personal needs and well-being to complete their education. The Chair of the department made it a priority that food insecurity would be a top priority in the department and across the institution.

According to Sabi (2019), negative perceptions of poverty, food aid, and overall food insecurity reflect a deep internal crisis of feeling shame that must be kept secret. Moreover, issues related to food insecurity and social stigma have been prevalent across the Global North and Global South respectively. Another university in South Africa documented nearly 60% of its student body experiencing high levels of food insecurity. In that same respect, students were also hesitant to apply for the "No Student Hungry Program," which was a university food security intervention program to help students experiencing hardships impacted by food insecurity. They believed that by participating in the program, their economic status would be exposed, further triggering stigmatization on and off campus. Also, the students felt that by not participating in the program, they would not be stereotyped as *food aid dependent*. Further, negative perceptions associated with the program appeared to demean the dignity of students instead of helping them with the common good. If they went to class hungry, they were reported to not perceive their lack of food as a problem. As with studies from the United States, the naming of food intervention programs was more harmful than the actual shame and embarrassment felt with receiving food aid. Surprisingly, about 39% of food in the United States is discarded, followed by 11% in Latin America, and 47% in Asia.

These figures represent a deeper problem at its core that individuals need to earn a living to afford food. Yet, the quality and composition of dietary diversity and food preferences coupled with lower wages impact the ability to sustain a quality food diet that impacts personal well-being and nutritional health.

College and university students might experience food insecurity for a variety of reasons, which include struggling to live on a budget as well as issues with managing a budget, food preparation skills, and planning for extraneous expenses (Zigmont et al., 2019). Some students have limited resources and are unable to cover their living expenses, while others might continue to live with their families in households experiencing food insecurity. Students experiencing food insecurity at any time in their academic careers have been found to have lower grades in school, to work while enrolled in classes, to receive financial aid, and to not have a reliable source of housing (Bruening et al., 2017), all which negatively affect educational attainment outcomes and experiences of advancing in adulthood.

In a 2015 survey of students experiencing food insecurity at an urban public research university, 80% of respondents reported that their experiences with food insecurity affected their class performance, and over 55% of those same students indicated that food insecurity hindered their ability to attend classes and participate in the course subjects. The everyday stresses associated with food insecurity affect how students succeed in school and manage their day-to-day activities. Food insecurity can affect a student's ability to complete their education, including poor health outcomes, low academic performance, and low cognitive functioning (Gundersen & Ziliak, 2015). Poor nutrition contributes to fatigue, difficulty with concentration, irritability, chronic diseases, and lower productivity. These factors directly affect a student's performance in the classroom, their retention and acquisition of knowledge, and graduation rates (Cady, 2014; Bruening et al., 2017; Shipley & Christopher, 2018).

Food insecurity among college students ranges anywhere between 18–59% (Gundersen et al. 2020; Buch et al., 2016). According to Cady (2014), students with food insecurity are more likely to receive federal nutrition assistance and need-based financial assistance. They are more likely to have experienced food insecurity before college entry than students with food security. A survey conducted by the Regents of the University of California (UC) found that students across the UC's ten campuses do not have access to adequate food. Students are also likely to continue skipping meals to save money (Martinez et al., 2016). In 2011, a survey conducted at the City University of New York revealed 40% of students reported being hungry, and 22% of surveyed students had skipped meals due to financial constraints.

In 2017, a student survey found that one in every ten undergraduate students at private research and land grant, sea-grant, and space-grant universities could not afford either one or all the following: food, course supplies, transportation, or professional attire for job interviews (Chao, 2017). Nationwide, 57% of students with food insecurity have been shown to have experienced food insecurity for the first time while in college. This suggests that students who are living on their own for the first time are likely to need financial, educational, and health-literacy training. Additionally, students might require supplemented information about food access resources, student financial aid, and managing schoolwork while on a budget (Cady, 2014) to manage their financial and personal health. These factors support the values and mission of the student success divisions on college campuses who prioritize students in all their objectives.

Evidence from low-income neighborhoods in Asia, Latin America, and Africa shows that fruit and vegetables are more affordable in public markets than at supermarkets. Huge volumes of fresh foods are supplied often direct-to-consumers, demonstrating high degrees of resilience and adaptability amidst market closures, they continue supplying food to school communities through the COVID-19 pandemic. Forty five percent of emergency food supply initiatives in Brazil were based on local supply chains. Patton-López et al. (2014) asked a sample of students if they had ever felt the need to compromise their health for the affordability of food. Student responses to their individual dining decisions included the following as the most critical factors in making food choices:

1. The menu selection available at an establishment, 60%
2. The price of food, 67%
3. The convenience of a location, 76%

Fifty-nine percent of students who were found to be food secure and used their campus meal plan were also found to have an inability to afford balanced, nutritional meals. These findings suggest that there is limited access to securing well-balanced meals on campus food establishments. Additionally, students' perceptions of the campus food environment underscore the importance of food affordability. It shows that, overwhelmingly, students will know that unhealthy food options are available but are uncertain about available healthy food options. The potential effects of insufficient dietary intake may include long-term health deficiencies, poor eating habits, susceptibility to various diseases, and social and intellectual development effects.

In another study of 30,000 students among 2-year and 4-year colleges and universities, over 50% of respondents at the 4-year universities reported that they had experienced or were currently experiencing food insecurity. Between 11–38% of students were experiencing or had experienced food insecurity at the 2-year colleges (Phillips et al., 2018). A different study examining college and university students in the state of Georgia found that approximately 29% of students experiencing food insecurity had poor psychosocial health and academic performance. The study's findings supported growing evidence for the high prevalence of food insecurity on college and university campuses and its effects on student health and success.

A lower GPA could predict the amount of time it takes to complete a degree. Students who have experienced severe food insecurity in college are more likely to have failed the courses that were part of their degree requirements (Freudenberg et al. 2013; Patton-López et al. 2014). Community college students who experience food insecurity have a lower likelihood of having a good GPA, which positions some of these students at potential risk of being unable to fulfill their degree requirements (Ilieva et al., 2019).

Students who report experiencing food insecurity are less likely to report a GPA of 3.1 or higher (Patton-López et al., 2014). Students with persistent structural, emotional, and physical barriers to accessing healthy foods may also experience impaired concentration and a lack of understanding of basic course concepts. Morris et al. (2016) found that some students who have experienced food insecurity in college have tended to have lower grade point averages (GPA). However, there are no recent studies to have found that students experiencing food insecurity are unlikely to graduate from college.

Food insecurity is an issue of global importance that affects many students in several countries, cities, colleges, and universities outside the U.S. (Mohamed & Ismail, 2022; Hanbazaza et al., 2016; Sabi et al., 2019; Theodoridis et al., 2018; Farahbaksh et al., 2015; Micevski et al., 2014). Food delivery to consumers involves many factors, including individuals and agencies. Within the food and agriculture systems, supply chains, market segmentation, and regulations across states and cities affect the composition of regional and local food systems (Nesheim et al., 2015). Significant events, such as the Great Recession and the COVID-19 pandemic, have changed the financial obligations and economic prospects of students entering institutes of higher education. Mangan et al. (2010) posited that the first step in alleviating food insecurity among students is to increase students' awareness of financial aid and other support services available to them on campus and in the community.

References

Bruening, M., Argo, K., Payne-Sturges, D., & Laska, M. N. (2017). The struggle is real: A systematic review of food insecurity on postsecondary education campuses. *Journal of the Academy of Nutrition and Dietetics, 117*(11), 1767–1791. https://doi.org/10.1016/j.jand.2017.05.022

Buch, K., Langley, S., Johnson, T., & Coleman, N. (2016). A university-community partnership to combat food insecurity among college students. *Partnerships: A Journal of Service-Learning and Civic Engagement, 7*(1), 16–26. http://libjournal.uncg.edu/prt/article/view/1332

Cady, C. L. (2014). Food insecurity as a student issue. *Journal of College and Character, 15*(4), 265–272. https://doi.org/10.1515/jcc-2014-0031

Cady, C. L., & White, C. C. (2018). Food pantries on campus to address student hunger. *New Directions for Community Colleges, 184*, 73–82. https://doi.org/10.1002/cc.20329

Chao, S. (2017, October 12). One in ten MIT undergrads can't afford food, survey finds. https://thetech.com/2017/10/12/case-survey-results

Daugherty, J. B., Birnbaum, M., & Clark, A. (2019). 'Having enough': Students' understanding of food insecurity and campus food pantry use. *Journal of Poverty, 23*(7), 600–620. https://doi.org/10.1080/10875549.2019.1652721

Davis, H., Sisson, S. B., & Clifton, S. (2020). A call for evidence to support food security interventions on college campuses. *Journal of American College Health*, 1–3. https://doi.org/10.1080/07448481.2019.1705829

Evans, B. A. (2016). Homeless and hungry in college. *Change: The Magazine of Higher Learning, 48*(1), 26–29. https://doi.org/10.1080/00091383.2016.1121082

Farahbakhsh, J., Ball, G. D., Farmer, A. P., Maximova, K., Hanbazaza, M., & Willows, N. D. (2015). How do student clients of a university-based food Bank cope with food insecurity? *Canadian Journal of Dietetic Practice and Research, 76*(4), 200–203. https://doi.org/10.3148/cjdpr-2015-020

Gundersen, C., & Ziliak, J. P. (2015). Food insecurity and health outcomes. *Health Affairs, 34*(11), 1830–1839. https://doi.org/10.1377/hlthaff.2015.0645

Gundersen, C., Hake, M., Dewey, A., & Engelhard, E. (2020). Food insecurity during COVID-19. *Applied Economic Perspectives and Policy, 43*(1), 153–161. https://doi.org/10.1002/aepp.13100

Hagedorn, R., & Olfert, M. (2018). Food insecurity and behavioral characteristics for academic success in young adults attending an Appalachian university. *Nutrients, 10*(3), 361. https://doi.org/10.3390/nu10030361

Hanbazaza, M., Ball, G. D., Farmer, A., Maximova, K., & Willows, N. D. (2016). Filling a need: Sociodemographic and educational characteristics among student clients of a university-based campus food bank. *Journal of Hunger & Environmental Nutrition, 11*(4), 569–577. https://doi.org/10.1080/19320248.2015.1128864

Ilieva, R. T., Ahmed, T., & Yan, A. (2019). Hungry minds: Investigating the food insecurity of minority community college students. *Journal of Public Affairs, 19*(3). https://doi.org/10.1002/pa.1891

Mangan, J., Hughes, A., & Slack, K. (2010). Student finance, information and decision-making. *Higher Education, 60*(5), 459–472. https://doi.org/10.1007/s10734-010-9309-7

Martinez, S. M., Maynard, K., & Ritchie, L. D. (2016). *Student food access and security study*. UC Nutrition Policy Institute. https://regents.universityofcalifornia.edu/regmeet/july16/e1attach.pdf

McArthur, L. H., Farris, A., Fasczewski, K., & Petrone, M. (2019). P139 student use and perceptions of a campus food pantry at Appalachian state university. *Journal of Nutrition Education and Behavior, 51*(7), S95. https://doi.org/10.1016/j.jneb.2019.05.515

McCarthy, C. (2019). Provide support to boost success of students facing food insecurity. *Student Affairs Today, 22*(5), 1–3. https://doi.org/10.1002/say.30636

Micevski, D. A., Thornton, L. E., & Brockington, S. (2014). Food insecurity among university students in Victoria: A pilot study. *Nutrition & Dietetics, 71*(4), 258–264. https://doi.org/10.1111/1747-0080.12097

Mohamed, H. S., & Ismail, K. A. (2022). Prevalence of food insecurity and associated factors among Cairo University students. *Journal of The Arab Society for Medical Research, 17*(1), 1–9. https://doi.org/10.4103/jasmr.jasmr_5_21

Morris, L. M., Smith, S., Davis, J., & Null, D. B. (2016). The prevalence of food security and insecurity among Illinois university students. *Journal of Nutrition Education and Behavior, 48*(6), 376–382. https://doi.org/10.1016/j.jneb.2016.03.013

Mukigi, D., & Brown, O. (2019). A text-delivered intervention to improve dietary habits, stress management behaviors and create awareness of food assistance resources among college students. *Journal of Nutrition Education and Behavior, 51*(7), S4–S5. https://doi.org/10.1016/j.jneb.2019.05.315

Nesheim, M. C., Oria, M., & Yih, P. T. (2015). *A framework for assessing effects of the food system.* National Academies Press.

Patton-López, M. M., López-Cevallos, D. F., Cancel-Tirado, D. I., & Vazquez, L. (2014). Prevalence and correlates of food insecurity among students attending a midsize rural university in Oregon. *Journal of Nutrition Education and Behavior, 46*(3), 209–214. https://doi.org/10.1016/j.jneb.2013.10.007

Phillips, E., McDaniel, A., & Croft, A. (2018). Food insecurity and academic disruption among college students. *Journal of Student Affairs Research and Practice, 55*(4), 353–372. https://doi.org/10.1080/19496591.2018.1470003

Sabi, S. C., Kolanisi, U., Siwela, M., & Naidoo, D. (2019). Students' vulnerability and perceptions of food insecurity at the University of KwaZulu-Natal. *South African Journal of Clinical Nutrition, 1-8.* https://doi.org/10.1080/16070658.2019.1600249

Shipley, G., & Christopher, M. (2018). Food insecurity on college campuses: Collateral damage of a societal crisis. *Journal of College and Character, 19*(4), 309–315. https://doi.org/10.1080/2194587X.2018.1517652

Silva, M. R., Kleinert, W. L., Sheppard, A. V., Cantrell, K. A., Freeman-Coppadge, D. J., Tsoy, E., Roberts, T., & Pearrow, M. (2017). The relationship between food security, housing stability, and school performance among college students in an urban university. *Journal of College Student Retention: Research, Theory & Practice, 19*(3), 284–299.

Theodoridis, X., Grammatikopoulou, M. G., Gkiouras, K., Papadopoulou, S. E., Agorastou, T., Gkika, I., Maraki, M. I., Dardavessis, T., & Chourdakis, M. (2018). Food insecurity and Mediterranean diet adherence among Greek university students. *Nutrition, Metabolism and Cardiovascular Diseases, 28*(5), 477–485. https://doi.org/10.1016/j.numecd.2018.02.007

Ullevig, S. L., Vasquez, L. L., Ratcliffe, L. G., Oswalt, S. B., Lee, N., & Lobitz, C. A. (2019). Establishing a campus garden and food pantry to address food insecurity: Lessons learned. *Journal of American College Health, 1-5.* https://doi.org/10.1080/07448481.2019.1705830

Zigmont, V., Linsmeier, A., & Gallup, P. (2019). Understanding the why of college student food insecurity. *Journal of Hunger & Environmental Nutrition, 16*(5), 595–610. https://doi.org/10.1080/19320248.2019.1701600

CHAPTER 2

Background of the Problem

Abstract Food insecurity in universities has emerged as a critical issue, affecting students' well-being and academic success. Factors contributing to food insecurity include rising tuition costs, inadequate financial aid, high living expenses, and limited access to affordable food options on or near campuses. Many students, particularly those from low-income backgrounds, face difficult trade-offs between paying for tuition, housing, and food. Additionally, the stigma surrounding food insecurity often prevents students from seeking assistance. This chapter explores the underlying causes of food insecurity in higher education, its impact on student health and academic performance, and the importance of addressing this issue to promote equity and success in academic environments.

Keywords Housing • Tuition • Food waste • Food desert • Economics

According to the United States Department of Agriculture, approximately 17 million individuals live in food deserts or regions lacking access to healthy foods and nutritious options and having limited access to a grocery store or supermarket carrying healthy, nutritious, and affordable foods. This figure represents college students with little or no income who have turned to food safety nets and aid groups while experiencing

S. Sharififard, *Community, Food Insecurity, and a Global Perspective on Campus Food Pantries*,
https://doi.org/10.1007/978-3-031-78223-7_2

challenges with attending school, finding employment, and participating in professional development opportunities. Students participating in school activities and extracurriculars while meeting the demands of a full course load must often choose between affording housing, tuition, and food. In some extreme cases, having limited or no access to food on or near campus threatens a student's academic performance and well-being.

Suppose a student decides to attend a school that considers their academic, social, and professional needs, such as career readiness and professional pathways. In that case, it is challenging to assess whether the campus neighborhood environment is in areas with a community for students to have a sense of identity and collegiality. To fully immerse oneself in the campus community, it is important to understand students' aspirations for attending a school that meets their needs to impact student performance in the classroom and at home. It is also essential to know how agricultural production is efficient while maintaining a healthy, desirable, and accessible food supply in local communities, especially college and university towns.

While current studies (Haskett et al., 2021; Henry, 2017; Raskind et al., 2019; Weaver et al., 2021) have attempted to explore the relationship between food security and success in college, others (Gundersen & Ziliak, 2018; Ruel et al., 2017; Wang et al., 2022) have explored the role of food safety nets in underserved communities. The global food system is highly complex and involves multiple actors and processes. The local food environment impacts food security and accessibility through:

1. the emergence of public and private food banks and food pantries
2. food waste and food policy and
3. food system activities related to food services and distribution

When studying students attending one urban and one suburban community college, 60% of the students surveyed at the urban community college had experienced some form of food insecurity, whereas 53% of the students at the suburban community college identified their experience with food insecurity (Maroto et al., 2015). The likelihood of this finding is that urban community colleges may have higher living expenses. Therefore, students who might have been financially dependent would be challenged to overcome such barriers during enrollment. Another likelihood is that more students enrolled in the suburban community colleges live with their families, where most outside expenses are used for school.

Further, the international students at both the urban and suburban community colleges had reported language barriers and limited transportation availability as a barrier to food access.

Food accessibility and its relationship to food and nutrient security are also relevant to communities representing high populations of college and university students. These factors differentiate the local food environment and affect the importance of the design of cities and neighborhoods, specifically, the benefits of structurally, economically, and geographically diverse food systems. Food products needed for food banks, food pantries, and food services have been limited and impacted by the COVID-19 pandemic (Beck & Gwilym, 2022). Food banks and food pantries often play a significant role in reducing the impact of food insecurity experienced by individuals. Similarly, food donations can frequently affect the types of food items needed or preferred by the community served. Some essential foods necessary for a household can become challenging to acquire as the supply of adequate, nutritious, and multiethnic foods, such as Halal, Jain, and Kosher, becomes less accessible and less affordable across neighborhoods and communities of food deserts.

Food deserts contribute to the need for more food in areas of limited macronutrient content. Also, the implications of the values portrayed by food programs, institutes of public higher education, and agricultural organizations can help bridge public-private sector partnerships, the convergence of public health concerns, and adverse effects on physical and mental health. While it is challenging to determine how diverse populations in urban, suburban, and rural areas would benefit from various food stores and food products, a lack of multiethnic food products in select food banks would limit improving the food environment overall, especially when there are limited or no viable alternatives.

Food providers vary across states and municipalities concerning the global food system (Beverly & Neill, 2022). Prolonged dietary intake consistent with nutritional recommendations affects the relationship between the environment and dietary patterns (Alsaffar, 2016). Recurring disparities in food environments further contribute to the need for healthful, nutritious, and specialized foods, such as organic and gluten-free. The lack of grocery and specialized food stores contributes to the problems of food and nutrient security problems or interest in current food consumption and practices. On average, the variety of convenience stores, easy-to-prepare, and ready-to-go snacks has increased, while the time spent on food preparation has decreased. Social influences and food behaviors of a community affect the purchasing behaviors of individuals in one area and their nutritional decisions (Leonard et al., 2014). The mix between secular and religious providers varies among communities (Minow, 2003).

The variety of convenience foods, easy-to-prepare foods, ready-to-go snacks, and consumption has increased, while the time spent on food preparation has decreased. However, technological innovations across the food system have allowed land-grant universities to work with farmers in creating safer and improved food for consumption (Food and Agriculture Organization of the United States, 2024). Over the past decade, schools have partnered with local farmers to introduce students to the food cycle and food manufacturing processes. Moreover, the food industry has partnered with land-grant universities to overcome the challenges associated with mass food preparation by innovation by reducing spoilage while improving the flavor of foods.

Consequently, Alston et al. (2008) argue that improving the quality of foods might increase food waste, leading to subsequent environmental problems. The demands for fresh and nutritious meals and the ease of disposing of food items have increased waste at some points of consumption. Meanwhile, policies that subsidize fewer goods can lead to lower micronutrient intake in response to having dietary diversity in the food system. Policies that support dietary diversity but fail to consider micro and macronutrient food quality can begin to restart a cycle of food insecurity among the population which can contribute to increasing the variety of food groups in a diet to help increase the essential nutrient intake and promote good health.

Further, social influences and food behaviors of a community affect grocery shopping behaviors of residents in that same area and their nutritional decisions. At the same time, genetic engineering techniques also have affected the overall food system. Ninety percent of all corn and soybean crops, the most abundant crops in the U.S., have been genetically engineered (Fernandez-Cornejo et al., 2014). The genetic engineering of such crops has affected the rate of food quality, food preferences, and food marketing. Different food products and the product assortment at food stores and other retailers can affect consumer interests.

Many universities have limited resources available for students who need food assistance. Some colleges and universities have launched campus food pantries as a solution to helping students with limited or no access to consistent food. A 2016 report studying over 3000 students across 34 college campuses indicated that only 14% of students with food insecurity had used a campus food pantry or a community food bank (Bruening et al., 2016). Rich-calorie, low-nutrient diets are linked to food and nutrient insecurity, and micronutrient deficiencies compromise physical and cognitive development. There are also several unintended consequences, such as increased food insecurity with the consumption of prepackaged, innutritious food, and chronic hunger.

Though the lack of sustainable campus food pantries has not been found within the specific context of higher education, it is part of the problem of perpetual food insecurity and poor health conditions. When students encounter challenges with securing their food needs from a campus food pantry, their other options consist of identifying foods outside the campus. Students can further find challenges with their necessary food security when the community does not offer options for their food needs. When preferred foods are not available at or near the campus, more challenges become prevalent in solving food insecurity issues.

A student's enrollment status and matriculation to a public or private institution do not necessarily affect the likelihood of experiencing food insecurity. Students enrolled part-time in four-year public colleges and universities have been significantly more likely to experience food insecurity (Blagg et al., 2017). Part-time enrollment could be attributed to several factors, including responsibilities for childcare, work, and the inability to attend school full-time, among other reasons. Having multiple responsibilities in addition to schoolwork can affect how students manage their time and take care of their health and wellbeing.

Further, students enrolled in technical and vocational education institutions have higher rates of food insecurity than students enrolled in traditional four-year colleges (Watson et al., 2022). This suggests that professional responsibilities might impact the path students take toward succeeding in their future careers. Professional responsibilities that are not related to academic success, or do not help with the program learning outcomes, also do not help students perform well as they advance in their academic careers.

Among 2-year and 4-year schools, unemployed students who are looking for work are the most likely to experience food insecurity (Blagg et al., 2017). Another study (Broton & Goldrick-Rab, 2016) researched 3000 students from low-income families enrolled in public two- or four-year college or universities in a state from the Midwest. Ninety percent of the surveyed students were concerned about not having enough money for their expenses to attend college for their first semester. The students reported having to change their shopping and eating habits and their driving and work schedules to accommodate their needs. Students also reported using credit cards more often, and either stopping or postponing paying their utility, medical, and cellular bills. Other changes in their activities included eliminating computers from their houses, reducing utility usage, and avoiding purchasing the required books and supplies for their academic courses.

In a university with an enrollment of approximately 16,000 students, 56% of undergraduate students with a median age of 22 had experienced some form of food insecurity, which included worrying about not having enough money for food. Twenty-six percent of respondents in the same study reported skipping meals due to a lack of money to buy food. Twenty-seven percent of the students said an inability to eat nutritious meals due to financial struggles; about 6.4% of participants reported severe food insecurity to the extent that they would not eat for a few days because of insufficient funds for food. As students begin school with distressed outcomes, it becomes increasingly difficult to foresee graduation prospects.

Many students often opt to leave college because of the stresses of supporting themselves while attending school. Family commitments, housing costs, accrued school-related expenses, and essential personal responsibilities have challenged even students who receive federal financial aid and work full time or part time. It is also difficult to succeed academically when financial aid and awarded scholarships do not adequately cover students' needs. These issues often cause students to take a leave of absence, preventing them from fully dedicating their time and commitment to academic success and schoolwork. Such situations may affect future career prospects and postgraduate opportunities.

Campus meal plans might not always serve as a viable option for preventing food insecurity. Many low-income students lack the funds necessary to afford a comprehensive meal plan and selecting a more cost-effective meal affects the number of meals a student may consume, which affects how much of nutritional value would be consumed. Low-income students are more likely to have jobs in which they are required to work for more than 20 hours per week to help pay for their college expenses. Students have reported having less time to prepare meals, which adds to the cycle of food insecurity. Further, students working for more than 20 hours per week are found to have lower GPAs than those who work fewer hours.

However, working for 10–15 hours per week has been shown to have positive impacts on a student's academic performance and learning engagement (Cady, 2014). If a student is working full time while enrolled in school on a full-time basis, they are required to choose between an academic course load that offers them the opportunity and necessary flexibility to fulfill their degree requirements; or consider advancing their curricular activities. While students might have the option to do both, they might also encounter compromising their food security because of participating in subsequent events related to school and higher learning engagement, socially and academically. These events and decisions can challenge prolonged success and the efficacy of student-faculty relationships.

Access to affordable, quality, fresh, and nutritious foods on campus depends on a student's location. The food environments at campuses can heavily contribute to poor eating patterns. The types of dining halls on campus and the high prices of healthy food retailers and eateries affect a student's access to good food (Marquis et al., 2019). Some students do not have meal plans, and others who live away from campus might not use the facilities, such as dining halls. Freshmen students may encounter significant transitions into independence and adulthood, which include behavioral, emotional, and social responsibilities and changes. Freshmen students also have reported higher levels of stress, higher rates of weight gain, and poor eating patterns (Cady, 2014).

A study at an urban university in Massachusetts studied students with food insecurity and their effect on academic performance. Food insecurity problems resulted in a higher risk of withdrawing or failing courses, which implies that students with food insecurity might not be completing their studies compared to students experiencing less food insecurity. Further, physiological and emotional distress incurred by completing a degree impacts how students react to their food insecurity needs and class performance.

The future consequences may become overwhelming if an individual does not begin to establish stable credit as a student. One's knowledge about financial, nutritional, and economic literacy could determine students' postgraduate outcomes and quality of life. The challenges faced with budgeting and financial management lead to persistent patterns of poor decision making that may affect one's credit score and budget, as well as their economic activity such as investing, spending, and saving. As a result, students will have fewer opportunities to compete for internships and job opportunities. Further, their potential to earn a degree and fulfill all responsibilities and school requirements becomes diminished.

Having a clear understanding of how specific college and university environments affect student wellbeing, academic achievement, and the motivation to succeed can help predict future economic factors such as poverty in adulthood, financial management, housing stability, and career success. The rising cost of attending college is attributed to the varying degrees of food insecurity, affecting many college students (Watson et al., 2022). Thus, many students may find themselves choosing between essential materials needed for attending college, such as books, and critical nutrients to sustain their concentration, energy, and wellbeing, like food.

Food pantry studies have been used to identify the increased concerns of student food insecurity; however, the foods available at food pantries have not been used to infer the relationships between food security and health food. The foods included in various diets and foods on campuses have not been examined from an administrative or leadership perspective. If students begin to strengthen their academic and personal foundations toward wellness early on, they will have a better understanding of the processes necessary to communicate their needs.

As campus leaders begin to recognize growth through advancing student success, campus food pantries will be better prepared to serve students who would not use a campus food pantry for their food security needs. Cultural and economic differences throughout campus are important in understanding some of the best ways to advance student success. Improving the function of the food pantry and involving students in relevant activities can increase participation in student affairs and student-driven work across campus. Other questions could begin to be raised among students when an important part of social and spiritual support is not available or recognized at the institutional level. These decisions are critical when factoring into the best kind of alternatives and policies for student success.

Students could begin to participate in a shared understanding of how a campus food pantry is used to sustain organizational goals and prepare for student wellness and wellbeing academically, socially, and emotionally. They can also consider how they can form positive relationships with individuals at all levels of the college or university and the effects of new student communications both directly and indirectly.

As students begin thinking about their program timeline and course completion, they might encounter difficulty in meeting their basic needs when they transition to advanced levels of their studies. Having these issues will lead students to incur additional costs to their education and will likely prolong their time in fulfilling their degree requirements. The costs associated with specific programs a student incurs would be important to assess if they would be willing to finance their expenses, such as food, with other financial obligations in pursuing a college degree. The campus food pantry's role becomes increasingly important for staff members to determine how or why their campus and food pantry effectively differentiate among food products. Their role also helps advise students who are fulfilling their degree requirements while encountering challenges with essential health needs.

Further, students are likely to schedule intermittent breaks on campuses where classes are spread across different buildings and locations; or if they are attending back-to-back classes that are not within the same or equal distance to one another. Consistent meal arrangements in between each class or participating in other curricular activities can be affected by those specific locations on campus, where one activity must be compromised for the other. The limited or restricted hours of operation even become more complicated for students when they cannot use the pantry at a time convenient for their schedules.

Available staffing and the limited capacity to manage facilities during holiday breaks and the summer reduce the ability to have the consistency of foods regularly. Therefore, a limited number of items can be offered in the pantry, but a student who needs those items will not be able to receive them at an appropriate time. The types of benefits a student can take advantage of can become significantly reduced, thereby the students would be expected to find other ways around campus or beyond school to obtain individual resources for their food insecurity needs. The relationships that pantry staff members develop with student patrons and the relations that faculty and staff form on campus grow the basis for partnering students with an active role on campus.

It is important to honor and respect all students' food choices as well as their individual dietary preferences. When a college or university successfully connects their institutional goals and values to the greater community, they can help other programs and projects to raise awareness about the persistent issues of food insecurity on campus. When students feel that their needs matter, they will have more space to communicate how their needs are reflected on campus. The opportunity to present campus issues and personal gaps in the learning environment can help educators understand the needs of their teaching and learning communities and consider how education barriers can reflect the circumstances facing the issues of their classes and institutions.

The direct relationships between the campus, the community, and the food pantry contribute to improved communication techniques between faculty, staff, and students. Also, being mindful and careful to student needs and understanding the structural, legal, and attitudinal barriers to food choices and food consumption is necessary for investigating food-related matters beyond colleges and universities. Such factors help to connect the relationships between higher education and the professional fields of business and law. Additionally, the various campus environments and the types of campus food pantries can show how an institution fulfills its mission, its role as an institution, and its overall student learning outcomes. For interventions to be successful or for the successful longevity of campus food pantries, sensitivity to cultural issues, as well as economic, social, and geographic barriers, should be visited.

From basic needs to psychosocial and academic health and wellness, the overall food quality is as significant to employ as the type of dietary guidelines and preferences a public institution or food vendor follows. To the same effect, higher costs for food items would affect the campus food pantries with limited operating budgets. In turn, the pantry staff would most likely continue to use external partnerships, such as food banks, to obtain their inventory. While it is significant to sustain such partnerships, both parties should have an advanced understanding of the populations they serve. Their objectives and common purpose could be used to identify the growing needs and emerging demands of new student populations and community concerns. Partnerships are an integral part of the college and university commitment to excellence and quality education.

Further, having the consideration to understand various student food preferences, organizational behaviors among the community, social characteristics, and neighborhood demographics can also help food retailers' prospects for entering a new market in a specific community. Similar considerations can assist higher education administrators in determining where the availability of nutritious, cultural, and healthful food is needed for a student population. Geographic, racial, and class disparities and health discrepancies can persist if the interplay between food preferences, social networks, and the shift of neighborhood characteristics over time does not align with specific student dietary preferences, food choices, and the availability of markets and food products. Using food pantries is just one solution to food insecurity. Food insecurity issues cannot be framed from a food scarcity lens, or the lens of food marketing.

If there are similar presented solutions, recognizing the importance of a solution or service such as a food pantry will be useful for many of the problems encountered with food insecurity. Different solutions with different purposes and interests can still reduce the meaning and effectiveness of the programs. A leader would not necessarily anticipate solving issues in a manner not reflective of the university, nor would they necessarily expect a temporary solution for most students. The amount of time and resources spent at an institution of higher education to help students fulfill their educational outcomes could show that students are pursuing their academics at a rate consistent with their peers and classmates.

Recently, the increasing changes in modern higher education and the types of needs required to sustain a quality education have changed the structure of the campus environment. With the growth in online degree programs and hybrid learning environments, food insecurity topics will continue to develop a greater need for public-private sector partnerships at the local level, which is important to advancing public policy discussions and urban planning initiatives. New communications are being used to enhance tech-related programs, and students can become contributors to their learning environments and personal growth. In terms of student success and wellness, leaders and administrators of colleges and universities could determine the extent to which their schools are responsible for providing a variety of basic needs to meet their students.

The critical work necessary for promoting student success outcomes and student learning objectives is a core component of academic performance indicators that help attain the student focus and the student's position in teaching and learning. There are also anticipated benefits for future subjects and society in general, which include:

- Better clarity around food systems and food environments.
- Making sense of the complexities associated with food security, food law, and food policy within the context of higher education.

References

Alsaffar, A. A. (2016). Sustainable diets: The interaction between food industry, nutrition, health and the environment. *Food Science and Technology International, 22*(2), 102–111. https://doi.org/10.1177/1082013215572029

Alston, J. M., Sumner, D. A., & Vosti, S. A. (2008). Farm subsidies and obesity in the United States: National evidence and international comparisons. *Food Policy, 33*(6), 470–479. https://doi.org/10.1016/j.foodpol.2008.05.008

Beck, D. J., & Gwilym, H. (2022). The food bank: A safety-net in place of welfare security in times of austerity and the Covid-19 crisis. *Social Policy and Society*, 1–17. https://doi.org/10.1017/S1474746421000907

Beverly, M., & Neill, C. L. (2022). Differences in food insecurity across the rural/urban spectrum—The role of trade flows. *Journal of the Agricultural and Applied Economics Association, 1*(1), 93–107. https://doi.org/10.1002/jaa2.11

Blagg, K., Gundersen, C., Schanzenbach, D. W., & Ziliak, J. P. (2017). Assessing food insecurity on campus. *The Urban Institute*. https://www.urban.org/research/publication/assessing-food-insecurity-campus

Broton & Goldrick-Rab. (2016). Working for college educational. *Evaluation and Policy Analysis 38*(3), 477–494. https://doi.org/10.3102/0162373716638440

Bruening, M., Brennhofer, S., van Woerden, I., Todd, M., & Laska, M. (2016). Factors related to the high rates of food insecurity among diverse, urban college freshmen. *Journal of the Academy of Nutrition and Dietetics, 116*(9), 1450–1457. https://doi.org/10.1016/j.jand.2016.04.004

Cady, C. L. (2014). Food insecurity as a student issue. *Journal of College and Character, 15*(4), 265–272. https://doi.org/10.1515/jcc-2014-0048

Fernandez-Cornejo, J., Wechsler, S., Livingston, M., & Mitchell, L. (2014). *Genetically engineered crops in the United States*. United States Department of Agriculture Economic Research Report. https://www.ers.usda.gov/publications/pub-details/?pubid=45182

Food and Agriculture Organization. (2024, July 6). Hunger numbers stubbornly high for three consecutive years as global crises deepen—UN report. https://www.fao.org/newsroom/detail/hunger-numbers-stubbornly-high-for-three-consecutive-years-as-global-crises-deepen%2D%2Dun-report/en

Gundersen, C., & Ziliak, J. P. (2018). Food insecurity research in the United States: Where we have been and where we need to go. *Applied Economic Perspectives and Policy, 40*(1), 119–135. https://doi.org/10.1093/aepp/ppx058

Haskett, M. E., Majumder, S., Kotter-Grühn, D., & Gutierrez, I. (2021). The role of university students' wellness in links between homelessness, food insecurity, and academic success. *Journal of Social Distress and Homelessness, 30*(1), 59–65. https://doi.org/10.1080/10530789.2020.1733815

Henry, L. (2017). Understanding food insecurity among college students: Experience, motivation, and local solutions. *Annals of Anthropological Practice, 41*(1), 6–19. https://doi.org/10.1111/napa.12108

Leonard, T., McKillop, C., Carson, J. A., & Shuval, K. (2014). Neighborhood effects on food consumption. *Journal of Behavioral and Experimental Economics, 51*, 99–113. https://doi.org/10.1016/j.socec.2014.04.002

Maroto, M. E., Snelling, A., & Linck, H. (2015). Food insecurity among community college students: Prevalence and association with grade point average. *Community College Journal of Research and Practice, 39*(6), 515–526. https://doi.org/10.1080/10668926.2013.850758

Marquis, M., Talbot, A., Sabourin, A., & Riopel, C. (2019). Exploring the environmental, personal and Behavioural factors as determinants for university students' food behaviour. *International Journal of Consumer Studies, 43*(1), 113–122. https://doi.org/10.1111/ijcs.12490

Minow, M. (2003). *Partners, not rivals: Privatization and the public good.* Beacon Press.

National Center for Education Statistics. (n.d.). *Status and trends in the education of racial and ethnic minorities.* http://nces.ed.gov/pubs2010/2010015/tables/table_24_4.asp

Raskind et al. (2019). Food insecurity psychosocial health and academic performance among college and university students in Georgia USA Abstract. *Public Health Nutrition, 22*(3), 476–485. https://doi.org/10.1017/S1368980018003439

Ruel, M. T., Garrett, J., Yosef, S., & Olivier, M. (2017). Urbanization, food security and nutrition. *Nutrition and Health in a Developing World*, 705–735. https://doi.org/10.1007/978-3-319-43739-2_32

Wang, J., Ding, X., Gao, H., & Fan, S. (2022). Reshaping food policy and governance to incentivize and empower disadvantaged groups for improving nutrition. *Nutrients, 14*(3), 648. https://doi.org/10.3390/nu14030648

Watson, M., Booth, S., Velardo, S., & Coveney, J. (2022). The orthodox and unorthodox food acquisition practices and coping strategies used by food insecure adults: A Scoping Review. *Journal of Hunger & Environmental Nutrition, 19*(6), 851–866. https://doi.org/10.1080/19320248.2021.2021121

Weaver, R. R., Hendricks, S. P., Vaughn, N. A., McPherson-Myers, P. E., Willis, S. L., & Terry, S. N. (2021). Obstacles to food security, food pantry use, and educational success among university students: A mixed methods approach. *Journal of American College Health.* https://doi.org/10.1080/07448481.2021.1873789

CHAPTER 3

Food Insecurity

Abstract In rural areas, factors such as limited access to grocery stores, transportation challenges, and economic hardships contribute to higher rates of food insecurity. These regions often lack adequate infrastructure and resources, resulting in food deserts where nutritious options are scarce. Conversely, urban areas, while more likely to have food outlets, experience food insecurity due to high living costs, income inequality, and systemic barriers affecting marginalized populations. This divide highlights the complex interplay between geography and access to food, emphasizing the need for targeted interventions. Addressing food insecurity requires a nuanced understanding of these disparities, as well as collaborative efforts to ensure equitable access to healthy food for all communities. The analysis suggests that current food pantries do not support macronutrient content. Food retailers surrounding college campuses also do not adequately support the dietary needs of students and the broader community. Additional consideration about the food source and where food comes from is essential for consumer behavior and neighborhood inclusivity in higher education and beyond.

Keywords Food insecurity • Urban • Rural • Suburban • Culture • Consumption • Mental health

S. Sharififard, *Community, Food Insecurity, and a Global Perspective on Campus Food Pantries*,
https://doi.org/10.1007/978-3-031-78223-7_3

According to the United States Department of Agriculture, food insecurity is categorized as either "low food security" or "very low food security." Low food security indicates that an individual has less quality and variety in their diet but is still obtaining enough calories. Very low food security is defined as eating less or not at all because of limited resources (USDA, 2018). Food systems policies have generally affected the common food supply and have contributed to some low food security among individuals and disproportionately influenced domestic and global food prices. While such aspects of food insecurity can be a combination of social, physical, traditional, and culture access, knowledge of marketing and legal incentives for food production is critical to understanding how such economic policies address low food security and caloric intake.

As the complex nature of the food supply chain evolves, the availability of food and access begins to affect the characteristics of the food environment. The individuals based in those communities also become affected, and not by coincidence. They can be troubled by the shifting changes in the environment which have not affected the community in ways that they are familiar with or recognize. Students who anticipated individual meal plans or expenses might have to change their anticipated commitments for other colleges that can support their goals and meet their financial means.

Food insecurity has been linked to structural deficiencies in the local food system, unpredictable circumstances, natural disasters, and chronic poverty. While food insecurity can be attributed to specific populations and geographic inequities, it has primarily been a public health issue. One of the first studies to have investigated food insecurity from a student perspective was conducted at a public research university in Hawaii (Chaparro et al., 2009). The researchers found that 45% of students surveyed were either experiencing food insecurity or at risk of experiencing food insecurity; and 95% of college students ($n = 441$) are dependent on their families for financial aid purposes. Fifty percent of those students had grown up in homes that struggled with food insecurity.

In the same study, 21% of students reported experiencing food insecurity, with 15% having low food security and 6% having very low security as defined by the USDA guidelines (Chaparro et al., 2009). It was suggested that students who lived off-campus with roommates, as well as students living on campus, were significantly more likely to experience food insecurity than students who lived with parents or relatives (Chaparro et al., 2009). Only 11% of students who lived with their parents were shown to have experienced food insecurity compared to 13% of students who lived off-campus with a spouse/partner.

Students' financial aid and living status would likely have been affected by expenses related to unpaid internships, volunteer experiences, and other professional activities. However, the study did not include first-year students in the survey, and the forms of debt accrual were not specified for upper-class students. Therefore, the participation of extracurricular activities would not have affected the types of acquired debt and perpetual debt in forthcoming studies as an upperclassman.

Researchers, Chaparro et al. (2009), Jones et al. (2018), Knol et al. (2017) Morris et al. (2016), have suggested that students who receive financial aid, and who live off-campus with no parents or relatives, may experience higher rates of food insecurity and higher financial strain.

In a study conducted at a public land-grant, space-grant, and sea-grant research university in Florida, researchers found that 31.52% of students were experiencing a form of food insecurity, with 42 respondents experiencing very low food security (El Zein, et al., 2018). Approximately 24.3% of students with food insecurity reported residing on-campus, and about half of the surveyed students, 49.7%, reported being employed in either a part- or full-time job.

The study did not report how long the students had held their positions. A qualitative sample of students who were obtaining employment for financial need and urgent food insecurity would be helpful to the study's findings. In the same study, 35% of the students had reported using student loans for their educational expenses. Almost 23% of respondents were Pell grant recipients who were experiencing the most food insecurity.

Overall, respondents in the sample were predominantly female, White, single, and undergraduate students. Undergraduate students were more likely to report food insecurity compared to graduate students.

One possibility of this finding could be that graduate students are better at budgeting their finances and are experienced with preparing consistent and balanced foods for their dietary needs. African-American/Black students were experiencing more food insecurity than White students by almost a 40-point percentage. The respondents indicated their status as international students had the highest rates of food insecurity compared to in-state and out-of-state students. This finding could have likely resulted from limited transportation, language barriers, and little familiarity with finding places to shop for food.

When asked if the students use the campus food pantry, most students reported to have been aware of the service, but only about 15% of the students had used the food pantry (El Zein et al., 2018). Students experiencing food insecurity may be hesitant to use the food pantry because of the stigma. By using the campus food pantry, students may feel embarrassed about their food insecurity situation and may avoid facing peers and colleagues who may find out about their experience. Additionally, students might not find the quality of the food available at the pantry suitable to their respective diets.

The study also examined the body mass index (BMI) of the student participants and their food acquisition but found no relationship between students' food insecurity and BMI categories and health outcomes. Physical fitness and wellness are essential to examine the effects of positive health with positive experiences in school, but measuring BMI is a flawed method for understanding food insecurity. A participant may have also recognized the study as a health and nutrition study for which body imagery was the primary data collection.

Other dietary, religious, ethnic, or personal food choices were not included. One student indicated that experiencing food insecurity for a long time caused him to purchase alternatives to healthier foods, such as cheap supplements, because it was hard to find quality nutritional food. Similar findings were also found by Huelskamp et al. (2019). Another student reported that federal financial aid only covers tuition, rent, and books, but food and gas were not included even with the support of a grant. Housing insecurity had also contributed to persistent issues. Participants were further asked how and where they acquired or obtained food, to which they indicated buying their groceries from low-cost chain grocers near their homes.

Students further reported transportation as another significant barrier to food access. Thus, it was more convenient to visit stores that were easier to access from their homes. When the students purchased food, they would typically buy less expensive food in bulk that would last for a more extended period. The same students recognized shared concerns with the quality of food with harmful health effects. Still, they were willing to acquire food from those stores for convenience and affordability.

Many participants also shared how they would participate in campus events that serve food. The foods were mostly snacks; and were not intended to replace a fully nutritious meal. Further research about the availability of foods in a food environment is important to understand how students with specific dietary needs can learn about the health content of their foods without risking their diets and the quality of the foods they routinely consume.

Besides lacking reliable transportation and a sufficient income, students indicated that working a full- or part-time job, attending classes, and studying contributed to a more considerable lack of time in their schedule. Shopping for food and preparing meals requires the necessary time needed to meet each need, and students expressed that it was common to opt to purchase foods that were convenient to prepare and were available at locations with short commute times.

Students who lived in the campus dormitories stated that their units often did not have proper or adequate kitchen appliances to cook and store food. These students faced challenges with purchasing, preparing, and storing food at home. Many participants indicated a willingness and preference to purchase healthy foods. However, finances and a lack of proper storage at home became a barrier. Additionally, the students experienced shame and stigmas with wanting assistance. This research suggests that students experiencing food insecurity have the desire to have consistent financial independence and maintain manageable food consumption habits without accessing food assistance. Instead, they were faced with several layers of economic, social, and political barriers.

Some students would seek grocery story jobs and foodservice industry employment so that there would be a possibility to access free or discounted food. Some other coping strategies students would also consume cheap, fast foods, share food with roommates, suppress their hunger with excessive fluid intake, and reduce their meal portions. Additional coping strategies included taking out loans, donating blood, taking online surveys for compensation, being research participants for studies, selling personal items, and attending different events with included meals whether they were invited.

Participants also admitted to stealing food to make ends meet by borrowing a friend or roommate's ID to use their available meal points. A few participants reported intentionally not paying bills to have money to pay for food. When participants were asked about their other coping strategies, female students indicated that they were more likely to use their networks. In contrast, male students indicated their willingness to go without food. The participants did not report for how long they would be willing to or have used such strategies, but most of the students indicated frequenting community soup kitchens if needed. For these students, it was a priority to earn their degree to secure a job after graduation and improve their quality of life. They also had considered the possibility that quitting school to work full time might be the solution to their current food insecurity issues; but not worth the long-term prospects of finding better employment opportunities.

According to Freudenberg et al. (2019), financially independent students are more likely to experience food insecurity than financially dependent students. While financial independence can advance one's position and status, adapting to adulthood and living individually for the first time may begin to form changes in diets, planning for meals, and food consumption, to secure their independence and investments. Students living on-campus are likely to have better diets than those living off-campus (Brunt et al., 2008). Students who live with their parents or relatives are less likely to experience food insecurity than students who live alone (Hughes et al., 2011). This finding is expected to support the idea that students in financially independent and financially dependent situations may perpetually face challenges paying for their college expenses.

Patton-López et al. (2014) found that 59% of students at a public university in Oregon had experienced food insecurity. The sample represented full-time undergraduate students within the 18–24 age range. Most of the students in the sample, 88%, reported having credit card debt, which may have been a factor of a student's food insecurity status. Twenty-nine percent of students experiencing food insecurity reported to have resided off-campus, and lived with roommates, lived alone, or lived with parents.

About 50.3% of students experiencing food insecurity indicated that they were employed in addition to attending college, working an average of 18 hours per week. The employment status, i.e., full or part time, is not reported. Still, students working to support themselves and access consistent sources of food are affected by time management skills the students have acquired for their academic course load, their work schedules, and setting time for food preparation. Food insecurity was not found to be significant by living arrangements, student enrollment status, health insurance status, physical activity, race, or other demographic factors.

Accruing debt, another area of concern for students (Gaines et al., 2014), persists when making payments for expenses with credit cards or loans (Hornak et al., 2010). According to a study conducted in 2006, 71% of college students charge books and food expenses to credit cards, and 84% of purchases charged to credit cards are primarily transacted by undergraduate students (McGlynn, 2006). Students with food insecurity,

Hispanic/Latino students, international students, and graduate students are more likely to use a credit card to purchase food. Students who are experiencing food insecurity are generally more likely to use a food pantry and borrow money to pay for college, compared to food secure students (El Zein et al., 2018). Students who borrow money to attend college have higher rates of food insecurity than students who do not borrow money for their educational expenses. Borrowing money and having more educational expenses could vary depending on the student's academic program or concentration. The degree-granting university may have different fees for its academic programs that cannot be generalizable from state to state.

According to Watson et al. (2017), students who are financially at risk are more likely to use credit cards for groceries and expenses. The researchers conducted a focus group interview using ten participants to observe food insecurity patterns and coping strategies. It was reported that the interview questions were validated by the institution's faculty, as well as an expert on food insecurity issues challenging students. The researchers used semi-structured interview questions that involved questions about food literacy and food security (Watson et al., 2017). The study informed participants about some of the terminologies in food security, which helps understand the perceptions and interpretations of the phenomenon and its social and academic implications. Students had the opportunity to share their experiences in a place that they might have considered a safe space. However, many generalizable assumptions can be made about the participants from the focus group.

For example, some students may have been curious about the issue of food insecurity or have been interested in the incentives offered by the researchers as compensation for participation. They also may have never been in or familiar with a situation involving food insecurity. Others might have participated as moral support for a friend. Their responses about food literacy would have been subject to biases as told by their observations, but not the perspective of a lived experience. During the study, the students may have withheld some of their opinions and beliefs to refrain from sharing any existing or perceived unpopular attitudes and beliefs.

Further, the students may have believed that their participation and responses would prompt further action or misleading judgment among the university or the participants. The departments, gender, year in school, and undergraduate/graduate student status were not confirmed in the study, which may have helped participants feel comfortable when the study would reflect the university as a case study, rather than individually assessing specific students and cohorts.

In a large study that assessed first-year undergraduate student nutrition, physical activity, and weight outcomes, researchers found that all students were required to purchase a meal plan if they lived in the residence halls. The survey for the study included items where the respondents were to indicate if the following statements were true:

> The food that I bought just didn't last, and I didn't have money to get more. I couldn't afford to eat balanced meals.
>
> In the past month, I had to cut the size of my meals or skip meals because there wasn't enough money for food.
>
> In the past month, I ate less than I felt I should because there wasn't enough money for food.
>
> In the past month, I was hungry but didn't eat because there was not enough money for food. (Bruening et al., 2018)

Any participant who gave affirmative answers to two or more questions was categorized as experiencing food insecurity. The authors also studied body composition, alcoholic consumption, and physical activity. Every participant was asked if they were a recipient of a grant from the Federal Pell Grant program. The results showed that meal patterns, perceptions of healthful eating, physical activity on campus, and mental health were all related to food insecurity. However, other dietary behaviors, weight status, and physical activity were not correlated with prior or recurrent food insecurity. The authors noted that these students' diets were deficient and that the differences in food intake could not be detected. The researchers also did not indicate some of the socio-demographics of the six residence halls in which this study sought to examine. The physical activities, class schedules, and internal functions of each community were not analyzed, which makes social desirability an issue for consideration. Results may also vary by the year of the students' enrollment in school, activities requiring their participation as part of their program's curriculum, and other campus characteristics. Further, the measures were mostly self-reported, which may have introduced biases. Finally, the researchers were limited to relationships of food insecurity with health behaviors and outcomes.

In a sample of 209 first-year undergraduate students attending a large university in the U.S., students experiencing food insecurity had significantly lower odds of eating breakfast, consuming home-cooked meals, and receiving food support from parents (Bruening et al., 2016). The students in the sample had also lived in the campus dorms. The researchers did not find significant differences between food insecurity and healthy eating habits on campus. There were no relationships identified among a student's status concerning weight, or stress. However, the researchers had found that students who rarely consumed breakfast, who rarely ate home-cooked meals, and who experienced higher levels of depression were significantly more likely to report food insecurity.

Further, the students who had been reported to consume fast foods often and who had unhealthy eating habits off-campus were more likely to have food insecurity experiences. Consuming regular healthy meals off-campus, however, was inversely related to food insecurity issues. Students were more likely to report food insecurity issues if their parents did not regularly assist with accessing food resources. The researchers further found that students overall had higher rates of anxiety and depression when experiencing food insecurity.

The researchers found no significance among food insecurity and healthy eating on campus, or consumption of unhealthy or fast foods, alcoholic consumption, and stress. Further, the questionnaire included items addressing students' food motivations and the determinants of students' eating behaviors among environmental, personal, and behavioral factors that are likely to influence university students overall.

The items addressed constructs concerning food insecurity, eating habits, culinary skills, environmental concerns, health concerns, mindfulness, sociability, and weight. Thirty-two percent of students who lived in residence halls or with a roommate expressed the understanding and importance of eating together. Of the respondents who indicated frequently cooking their meals, 33% lived with roommates, whereas 51% of the respondents who indicated that they do not usually prepare their meals lived in residence halls. Among the students who reported selecting foods based on price, 34% of those living with a roommate were more price-sensitive, and those who lived by themselves either in an apartment or in a residence hall were less represented. Of the students who were less interested in trying new foods or reported less variety in their diet, 41% lived in residence halls.

Relationships were found between food insecurity, place of residence/ living arrangement, and reported a level of food insecurity, living with a roommate, and living alone. Of those unconcerned by food insecurity, 33% were living in residence halls, and 30% were living with roommates

(Bruening et al., 2016). Among adult students, food insecurity was related to a higher risk of poor dietary quality, binge eating, chronic disease, and lower work productivity. The prevalence of this study found anxiety and depression to be as high among first-year undergraduate students experiencing food insecurity compared to their peers. The combination of food insecurity and public health issues might affect learning and attitudinal success, which is an area to further explore in higher education.

A different study found that among the prevalence of low and very low food security, college/university type would vary. As such, 23% of private college/university students, 33% of Historically Black Colleges and Universities, 34% of public university students and 37% of technical and vocational college students were found to experience food insecurity, with 39% of the students found to have been very low in food security (Raskind et al., 2019). Students who were not employed were less likely to experience food insecurity. Unemployed students and students living off-campus had a higher likelihood of experiencing food insecurity than those who were living in university housing. The more a student experiences food insecurity, the more likely they are to experience an increase in depression and anxiety, and a decrease in their overall GPA.

The students who live on their while enrolled in college and who prepare, plan, and purchase their foods might experience differing levels of food security among their peers. Students who lack cooking and preparation skills for meals may also opt to purchase costlier or convenience foods (Gaines et al., 2014). A sample of students living off-campus while attending a public research university in Alabama found that 38.3% of students were experiencing food insecurity and had lower cooking self-efficacy and food preparation skills than students who were not experiencing food insecurity. The researchers reported that food insecurity status, coupled with financial aid debt, contributed most to food preparation behaviors among more males than females (Knol et al., 2019). Using a four-point Likert Scale, with (1) indicating very inadequate and (4) very adequate, respondents were asked to rate their skills from cooking to selection and supply of cooking appliances, time management, and financial resources. Also, using a five-point Likert Scale, with (1) indicating not at all confident and (5) extremely confident, respondents were asked to rate their level of confidence in completing tasks such as cooking a nutritious meal, following a recipe, time management, and level of cooking skills.

Most students reported high food security, although about 20% indicated experiencing anxiety about their food supply. The researchers also assessed selected aspects of food insecurity among community college students asked students at two colleges:

1. If their households ever ran out of money to buy food.
2. If they cut meals or skipped meals because of a lack of money.
3. If they used government or emergency food assistance programs.

The researchers found that between 24–27% of students lived in households that ran out of money to buy food; 22–24% reported cutting or skipping meals because of running out of cash; 18–28% used government-assisted food programs; 7–10% reported using emergency food.

The results in the study did not show whether students who lived in university dormitories and who had never cooked a personal meal were at a higher risk of developing food insecurity than those who cook often.

The authors found no relationship between demographic characteristics and food insecurity status; however, the research could contribute further to explaining how financial aid debt varies across the different demographic representation. Additionally, the authors would better help other researchers understand how food preparation changes or continues to improve from the beginning of the academic year, and each year after. Finally, the researchers did not indicate if the students' confidence levels changed as their cooking skills improved, or if their appliances and time management converged with purchasing better supplies and tools to prepare food as their time management improved.

Food insecurity has also affected students who are employed, participate in a campus meal plan, or seek other financial assistance. In a survey conducted by the Wisconsin HOPE Lab, now the Hope Center at Temple University in Philadelphia, Pennsylvania, researchers found that among low- and moderate-income undergraduate students, 77% of those who lived at home with parents would provide monetary support to their families.

Having primary responsibilities for their family members added to the issue of food insecurity, which led to their inability to obtain financial independence. Other surveyed students had parents who would either purchase groceries or provide them with money for food. Students with food insecurity reported receiving food-related public assistance. However, the number of students experiencing food insecurity and receiving assistance was higher than students who were receiving public assistance and were food secure (Broton & Goldrick-Rab, 2018).

Sixty-four percent of students experiencing food insecurity reported also experiencing housing insecurity and believed that hunger and housing problems affected their education. Students cited consequences such as missing classes, dropping classes, and not purchasing required textbooks. Forty-three percent of the students with meal plans reportedly experienced food insecurity and insufficient food to eat. Some of these students reported not eating for a day to have enough points to last for the academic term. Some respondents indicated that they had received free food in the past.

In a different study (Broton et al., 2018) conducted by the same lab, the researchers surveyed students from community colleges. The researchers found no strong relationships between community college student responses to food and housing insecurity (Broton et al., 2018). It is unknown if any of the students had transferred to four-year universities, or if there was significant retention among students experiencing food insecurity during their first semester. The participants' programs could also be studied for further analysis of the rates in which individual students enrolled in different programs experience varying levels of low or very low food security. In this study, there is no indication if the students experiencing food insecurity were more likely to commute from far distances, or if their locations were near a grocer or food store. Further, the researchers did not suggest a relationship between food insecurity and academic performance. Racial demographics were also not found to have corresponded with food insecurity in the study.

References

Broton, K. M., & Goldrick-Rab, S. (2018). Going without: An exploration of food and housing insecurity among undergraduates. *Educational Researcher, 47*(2), 121–133. https://doi.org/10.3102/0013189X17741303

Broton, K., Weaver, K., & Mai, M. (2018). Hunger in higher education: Experiences and correlates of food insecurity among Wisconsin undergraduates from low-income families. *Social Sciences, 7*(10), 179. https://doi.org/10.3390/socsci7100179

Bruening, M., Brennhofer, S., van Woerden, I., Todd, M., & Laska, M. (2016). Factors related to the high rates of food insecurity among diverse, urban college freshmen. *Journal of the Academy of Nutrition and Dietetics, 116*(9), 1450–1457. https://doi.org/10.1016/j.jand.2016.04.004

Bruening, M., van Woerden, I., Todd, M., & Laska, M. N. (2018). Hungry to learn: The prevalence and effects of food insecurity on health behaviors and outcomes over time among a diverse sample of university freshmen. *International Journal of Behavioral Nutrition and Physical Activity, 15*(1), 9. https://doi.org/10.1186/s12966-018-0647-7

Brunt, A., Rhee, Y., & Zhong, L. (2008). Differences in dietary patterns among college students according to body mass index. *Journal of American College Health, 56*(6), 629–634. https://doi.org/10.3200/JACH.56.6.629-634

Chaparro, M. P., Zaghloul, S. S., Holck, P., & Dobbs, J. (2009). Food insecurity prevalence among college students at the University of Hawai'i at Mānoa. *Public Health Nutrition, 12*(11), 2097–2103. https://doi.org/10.1017/S1368980009990735

El Zein, A., Mathews, A., House, L., & Shelnutt, K. (2018). Why are hungry college students not seeking help? Predictors of and barriers to using an on-campus food pantry. *Nutrients, 10*(9), 1163. https://doi.org/10.3390/nu10091163

Freudenberg, N., Goldrick-Rab, S., Poppendieck, J. (2019). College students and SNAP: The new face of food insecurity in the United States. *American Journal of Public Health, 109*(12), 1652–1658, (December 1, 2019).

Gaines, A., Robb, C. A., Knol, L. L., & Sickler, S. (2014). Examining the role of financial factors, resources and skills in predicting food security status among college students. *International Journal of Consumer Studies, 38*(4), 374–384. https://doi.org/10.1111/ijcs.12110

Hornak, A. M., Farrell, P. L., & Jackson, N. J. (2010). Making it (or not) on a dime in college: Implications for practice. *Journal of College Student Development, 51*(5), 481–495. https://doi.org/10.1353/csd.2010.0003

Huelskamp, A., Waity, J., & Russell, J. (2019). Effects of campus food insecurity on obesogenic behaviors in college students. *Journal of American College Health*, 1–4. https://doi.org/10.1080/07448481.2019.1684298

Hughes, R., Serebryanikova, I., Donaldson, K., & Leveritt, M. (2011). Student food insecurity: The skeleton in the university closet. *Nutrition & Dietetics, 68*(1), 27–32. https://doi.org/10.1111/j.1747-0080.2010.01496.x

Jones, P. J., Park, S. Y., & Lefevor, G. T. (2018). Contemporary college student anxiety: The role of academic distress, financial stress, and support. *Journal of College Counseling, 21*(3), 252–264. https://doi.org/10.1002/jocc.12107

Knol, L. L., Robb, C. A., McKinley, E. M., & Wood, M. (2017). Food insecurity, self-rated health, and obesity among college students. *American Journal of Health Education, 48*(4), 248–255. https://doi.org/10.1080/19325037.2017.1316689

Knol, L. L., Robb, C. A., McKinley, E. M., & Wood, M. (2019). Very low food security status is related to lower cooking self-efficacy and less frequent food preparation behaviors among college students. *Journal of Nutrition Education and Behavior, 51*(3), 357–363. https://doi.org/10.1016/j.jneb.2018.10.009

Morris, L. M., Smith, S., Davis, J., & Null, D. B. (2016). The prevalence of food security and insecurity among Illinois university students. *Journal of Nutrition Education and Behavior, 48*(6), 376–382. https://doi.org/10.1016/j.jneb.2016.03.013

McGlynn, A. P. (2006). College on credit has kids dropping out. *The Education Digest, 71*(8), 57–60. https://eric.ed.gov/?id=EJ741064

Patton-López, M. M., López-Cevallos, D. F., Cancel-Tirado, D. I., & Vazquez, L. (2014). Prevalence and correlates of food insecurity among students attending a midsize rural university in Oregon. *Journal of nutrition education and behavior, 46*(3), 209–214.

Raskind, I. G., Haardörfer, R., & Berg, C. J. (2019). Food insecurity, psychosocial health and academic performance among college and university students in Georgia, USA. *Public Health Nutrition, 22*(3), 476–485. https://doi.org/10.1017/S1368980018003439

U.S. Department of Agriculture, Economic Research Service. (n.d.). Food security in the U.S. U.S. Department of Agriculture. https://www.ers.usda.gov/topics/food-nutrition-assistance/food-security-in-the-u-s/

Watson, T., Malan, H., Glik, D., & Martinez, S. (2017). College students identify university support for basic needs and life skills as key ingredient in addressing food insecurity on campus. *California Agriculture, 71*(3), 130–138. https://doi.org/10.3733/ca.2017a0023

CHAPTER 4

Factors Contributing to the Growth of Campus Food Pantries

Abstract The growth of food pantries on college campuses has emerged as a vital response to rising food insecurity among students. These pantries provide essential support and foster community and a sense of belonging. However, there is an increasing recognition of the need for multiethnic foods in these services to better serve diverse student populations. Many students come from varied cultural backgrounds and dietary preferences, and access to familiar, culturally appropriate foods can significantly impact their overall wellbeing and academic performance. By expanding food pantry offerings to include a range of ethnic foods, campuses can create more inclusive environments, ensuring that all students feel valued and supported. This approach not only addresses food insecurity but also enriches campus culture and diversity.

Keywords Food banks • Dietary preference • Culture • Real estate • Multiethnic • Community

Food insecurity is as problematic and concerning on college campuses as housing insecurity, where students experience levels of homelessness or disruption in reliable housing. Pantry staff members in the study reported informing other campus officials about ways to engage in activities related to supporting student success and meeting basic needs on campus. One

S. Sharififard, *Community, Food Insecurity, and a Global Perspective on Campus Food Pantries*,
https://doi.org/10.1007/978-3-031-78223-7_4

interesting difference found from the results of this study included the number of student patrons visiting the campus food pantries. The number of student patrons has been found to have increased overall, with large increases in the number of total unique or first-time visitors. However, these figures overall do not include staff and faculty who also use their campus food pantry's services.

Because some of the food pantries are partners of local food banks, it is reasonable to suggest that certain campus food pantries might not have primarily been designed with the intent to allocate specific student food needs. Because food-focused interventions have unintended consequences, the leaders across institutes of higher education could recognize the structures and shifts in their neighborhood food environment and further develop an understanding of students' cultural and dietary preferences. Students who practice a dietary regimen that may alter within the school year, such as athletes, or arts and entertainment performers, may find pantry services useful for some of their short-term or occasional needs.

Factors affecting the growth of campus food pantries are different across campuses and neighborhoods. Participants had indicated that most of the campus food pantries overall are in areas where many students visit or take classes throughout the day. There are challenges with operating food pantries in some specific areas on campus, as also described in previous literature. Some of the campus food pantries are not always located in a place where they were initially planned to be present. Also, their location often changes within an academic year. Frequent communications are necessary to ensure that updates to the pantry's services and operations are communicated with students.

Further, the accessibility of food destinations from a student's location is important to consider as students change schedules from term to term and year to year while fulfilling degree/course requirements. Students can be affected by their ability to find a food pantry when changing their academic and work schedules. Students who previously had access to their food pantry services may not have a consistent method of visiting the pantry as they might have had. Such changes can also affect the hours of operation a campus designates for their food pantries or when deciding to open more food pantries across campus to reach more students.

Areas with a larger concentration of several convenience stores may not necessarily be places where developers choose to invest. However, it would be an opportunity to consider, as colleges and universities have a different market that is important across communities. In that same respect, future developers evaluate implications of the values that a public institution of higher education serves, bridging public-private sector partnerships. Additionally, high start-up costs and the availability of attractive land for developers to build grocery stores with more product offerings near college campuses are also considerations for professionals, such as realtors.

Recurring disparities in food environments contribute to the lack of healthful, nutritious, and specialized foods. The lack of grocery stores, coupled with a lack of specialized food stores, each contributed to the problems of food insecurity differently. Economically, if convenience stores were to offer more food items, including specialized and healthy food, they would likely charge more for them, which would not necessarily be affordable for all students.

More than half of the sample representatives indicated that their food pantry is in a location with the highest number of student gatherings, such as the student center, where students socialize and get information about student services or participate in programs and cultural activities. According to findings, students who have needed to use the campus food pantry have frequently not known the pantry's location. Other students have not understood the food pantry's purpose and primary use. The absence of sidewalks as a barrier to the physical environment of the food environment has also been a point of consideration.

Yet, the factors that differentiate the food environment from the overall campus environment include the importance of the design of cities and neighborhoods. In contrast to the literature, most students who visited the food pantry were not found to have encountered issues specifically with accessing or locating the pantry because of physical space on campus. Overall, access to food by distance did not heavily influence the likelihood or willingness of students visiting their campus food pantry. However, a local grocery store or food pantry located close to a student's residence or dorm might not be of great utility to a student.

As with neighborhoods described in the case study analysis, many students would have to walk a longer distance or use a different mode of transportation to arrive at a grocer or food retailer that met their specific food needs beyond campus. The same would apply for the campuses located in communities that typically lack safe and accessible crosswalks and traffic lights. Students who continue to use their campus food pantry could consider how another service on campus would better meet their basic needs if transportation and accessibility are of greater significance to their travel and class scheduling needs. Most significantly form a policy perspective, the lack of transportation also affects students from traveling outside their campus, which requires the necessity to have more on-campus food options for students to acquire items they need to sustain their diets, with extended hours of operation to meet the demand.

The food pantries help solve transportation barriers for students who reside on campus, but only in limited capacities. Fully online students, commuter students, and students with limited or no vehicle or transportation access encounter different restrictions in visiting their campus food pantry. Concerning spatial complexity and the effects on student accessibility, the food pantries are not always open when students participate in classes. If they were to use the food pantry, they would have to adjust their schedules or find other ways to meet their food security needs while commuting or receiving assistance from peers on campus.

Episodic cash flow is another factor contributing to the growth of campus food pantries. It is one of the more significant problems affecting students, rather than the direct effects of extreme forms of poverty and hunger. Students who do not have consistent forms of income or employment have difficulty with securing their food needs and academic expenses. Because of the uncertainties of having stable cash flow, food expenses are part of a more complex conversation surrounding tuition and student services, and the cost of attending school. The cost of maintaining campus food pantries is also an important factor in the process of opening and maintaining services that are ready to meet the needs of students.

Based on these additional findings, campus food pantries will likely continue to increase throughout colleges and universities to supplement some of the basic needs that are not available in many neighborhoods overall.

Campus Food Pantries and Students' Food Insecurity

Patrons might have specific food and dietary needs. The campus food pantry might encounter several challenges in consistently meeting the demand of providing every food product type that might be needed or requested on campus. There were pantries in this study that had been identified or described as an up-and-coming project created to meet immediate and recurring basic student needs, others were launched to promote student wellbeing and health. Generally, the rules, regulations, laws, and guidelines used to operate a campus food pantry affect the types of foods and items a pantry can offer in an academic year.

It is challenging to determine how student populations would benefit from a variety of food stores and food products on campus as well as within the neighborhoods of their residence. A lack of multiethnic food products in the food pantries, while not necessarily found to help improve the food environment overall, does not imply that the essential quality of foods in low-income, high-poverty neighborhoods are scarce. Multiethnic representation in the U.S. varies significantly from region to region.

Students could potentially visit their campus food pantry that is either entirely or partially devoid of dietary foods, which can be more difficult when deciding to visit their campus food pantry. The food pantry can differentiate itself from the local or community food bank, making it more connected to student and campus needs. Further, one cannot assert that students would choose healthful, nutrient-rich foods if provided with adequate access. Also, the availability of specific foods in a neighborhood does not necessarily infer an increased or decreased consumption of healthful food items. However, as it pertains to findings within the literature (Nazmi et al., 2018, Martin et al., 2021, Ginsburg et al., 2019, Bryan et al., 2019), there are further implications for considering the population's health disparities.

As it relates to the spatial complexity of food insecurity, neighborhoods with a lower variety of foods would not be as important for a student experiencing food insecurity if there is no access to the specific food item(s) they need. Additionally, students might enroll in a university that is in a neighborhood with a more extensive selection of restaurants. Based on these findings, it is likely that food insecurity will continue to increase on college campuses unless the quality and affordability of food stores are better than the options currently available to students.

Also, food pantries on campus require students to show a form of student identification to use the pantry. If a student misplaces their student identification or forgets to bring a form of identification, it might delay the process and prevent them from continuing to use the pantry when basic instructions and eligibility are not carefully followed. Additionally, students might not use the service if they feel that they are being traced or monitored for every visit. Students might wish to use their campus food pantry confidentially. The process of tracking students who frequently use the pantry and the recommendation of having students apply for benefits can also further increase the stigma surrounding food insecurity.

Students may feel that their needs will not be adequately addressed by submitting a basic application form. Also, having to wait for a certain amount of time to have an application approved is another challenge that would prompt a student to reconsider a trip to the pantry or appointment with a campus Case Manager. A student might also feel shame in discussing their concerns with a Case Manager. As shown in the study's findings, many interns and student workers manage the campus food pantries' operations. Patrons might not be willing to meet with someone among their peer group who would be responsible for assisting the student with completing a basic needs application.

Campus Food Pantries and Students' Dietary Needs

If a local food bank serves as the only provider to supply food items that are not an option for consumption to a group of students, those same individuals might feel cultural or social isolation. In turn, students might consume smaller portions of food that are not a part of their regular dietary intake or use food items that are not available. If the requirements for maintaining a balanced and healthy level of dietary intake are excluded from a student's respective dietary intake, the students might not have the opportunity to seek food from other food retailers or campus services (Hwalla et al., 2016).

Further, the more that campus food pantries rely on the partnership or services and donations offered by their community food banks, the more likely it is for some issues surrounding food insecurity to persist. Many essential foods necessary for large groups of students will also be more challenging to acquire as the supply of adequate, nutritious, multiethnic, and healthy foods becomes less accessible around college and university campuses and more difficult to find in the neighborhood. If students feel excluded from the services and food items that their campus pantry offers, the number of student patrons could decline. A decline in pantry services would not imply less food insecurity around campus. Instead, the decline would challenge the effectiveness of food pantries on campus.

Further, a campus could not trace every type of dietary preference, custom or desired food rule. Also, some students might not follow religious or dietary customs, irrespective of a faith-based, athletic, performance, or physical training background. Still, it is beneficial to understand how all members of the student body access and consume food and determine how foods are acquired, prepared, stored, and handled. One cannot find what they are looking for in terms of food, without compromising the other, especially regarding dietary customs and food preparation. The spatial complexity of the campus food environment can, in many ways, inform how the community food environment contributes to better and preferable consumer interests over time.

Additionally, dietary preferences and food laws might be more conducive to secular food rules or preferences when operating a campus food pantry that serves a diverse student body. However, it is one additional consideration to infer what happens when religious/secular relationships begin to cause shifts in public institutions. The mix between secular and religious providers varies among communities (Minow, 2003). Food providers also vary across states and municipalities concerning the global food system. The food system changes in response to consumer demands and is affected by individuals who follow specific food customs over time.

If campus food pantries were to introduce more food products derived from dietary laws or cultural and customary contents, economic, political, and social changes might occur and affect some student preferences, behaviors, and practices. Food distributors and foodservice providers would also see the effects of their business models and individual contracts with institutions in providing a variety of foods that would also be affordable. As such, public institutions of higher education that would prefer to keep foods of a religious or cultural purpose separate from their campus food pantry or campus culture, in general, could find other viable solutions to providing relevant and inclusive food products for the populations they serve.

References

Bryan, A. D., Ginsburg, Z. A., Rubinstein, E. B., Frankel, H. J., Maroko, A. R., Schechter, C. B., Stowers, K. C., & Lucan, S. C. (2019). Foods and drinks available from urban food pantries: Nutritional quality by item type, sourcing, and distribution method. *Journal of Community Health, 44*(2), 339–364. https://doi.org/10.1007/s10900-018-0592-z

Ginsburg, Z. A., Bryan, A. D., Rubinstein, E. B., Frankel, H. J., Maroko, A. R., Schechter, C. B., Stowers, K. C., & Lucan, S. C. (2019). Unreliable and difficult-to-access food for those in need: A qualitative and quantitative study of urban food pantries. *Journal of Community Health, 44*(1), 16–31. https://doi.org/10.1007/s10900-018-0549-2

Hwalla, N., El Labban, S., & Bahn, R. A. (2016). Nutrition security is an integral component of food security. *Frontiers in Life Science, 9*(3), 167–172. https://doi.org/10.1080/21553769.2016.1209133

Martin, K., Xu, R., & Schwartz, M. B. (2021). Food pantries select healthier foods after nutrition information is available on their food bank's ordering platform. *Public Health Nutrition, 24*(15), 5066–5073. https://doi.org/10.1017/S1368980020004814

Minow, M. (2003). Public and private partnerships: Accounting for the new religion. *Harvard Law Review, 116*(5), 1229–1270.

Nazmi, A., Martinez, S., Byrd, A., Robinson, D., Bianco, S., Maguire, J., Crutchfield, R. M., Condron, K., & Ritchie, L. (2018). A systematic review of food insecurity among US students in higher education. *Journal of Hunger & Environmental Nutrition*, 1–16. https://doi.org/10.1080/19320248.2018.1484316

CHAPTER 5

Post COVID-19 Food Insecurity Discussions on Campus

Abstract Food insecurity is increasingly recognized as a critical issue within the framework of diversity, equity, and inclusion (DEI) in educational and community settings. Discussions around food insecurity reveal disparities that disproportionately affect marginalized groups, including low-income students, racial and ethnic minorities, and first-generation college students. These populations often face systemic barriers that limit their access to nutritious food, exacerbating health disparities and hindering academic success. Integrating food security into DEI initiatives highlights the need for holistic approaches that address these inequities. Promoting equitable access to resources, culturally relevant food options, and supportive services fosters an inclusive environment where all individuals can thrive. By prioritizing food security in DEI discussions, institutions of higher education can work toward creating a more just and equitable society for all.

Keywords Diversity • Equity • Inclusion • Justice • Welfare • Security • Belonging

In 2023, 87% of participants surveyed indicated that food insecurity is a topic of discussion among faculty and staff at their campus post COVID-19. Nineteen participants reflected that an increase in basic needs was met by

S. Sharififard, *Community, Food Insecurity, and a Global Perspective on Campus Food Pantries*,
https://doi.org/10.1007/978-3-031-78223-7_5

first year students, fully online students, and first year graduate students. Eleven participants also noted similar findings but shared that in internal student success surveys distributed on campus, food insecurity policies have begun to take place. For example, as part of DEI-related initiatives, higher ed administrators are including inclusive language in their Bylaws and code of conduct that reflects a respect for food, basic rights, dignity, and diversity of cultures. Traditional holidays that were not officially recognized by legislatures are included on campus traditions and student affairs. During the pandemic, student leadership groups from a variety of campus departments and activities answered the class to leadership and began hosting virtual and on-campus events to include their backgrounds and representations in a safe space. As a student worker participant noted

> COVID allowed us to have events that we normally did not have the money to have on campus. Now, we can be even more inclusive and invite students and non-students to come to our events, understand our cultures and traditions, and have a deeper conversation for the community we are striving to become. (Participant, 26)

Others noted the relationship between food and society and food and place to bring inclusivity to their campuses.

> We had an event in 2022 that was about food and agriculture. We thought that only 5 friends would show up. We ended up having over 60 people join us including our Dean, a member of the President's staff, and a lot of our professors. This showed us that there is interest on campus, and we are addressing the problems of food insecurity at a higher level. We can't have band-aid solutions that temporarily solve the issue without going deep into the intersectional conversations that we need to have long-term. (Participant, 12)

> Not only did COVID give us a chance to change the way a hungry student looks like, but we had faculty come out, show up, and show us their support. They saw what we were doing, and they got involved however they could. Whether it was going to a city council meeting, or starting a food policy group, we never left it out of discussion. (Participant 16)

As it relates to international students, a coordinator of student success in France mentioned,

> In France, food is a very adventurous topic that should be explored and made meaningful. We tell stories about food. We want to share experiences and memories. Students would come to the classroom and make bread, and then share a story about their home. With the pandemic, the art of gastronomy is lost. It doesn't matter how many times you went to the café and had a meal with your friends or were part of a lecture in class, we now gather in homes, and it is up to the next generation to have those conversations on food and place. We need to have more stories, and we need to have the preservation of traditions fluent in our activities. (Participant, 18)

> Food insecurity has affected our international students. Students walked into my office at 10pm asking if there was any food leftover. It breaks my heart that we haven't had these discussions at length, but we now have a task force looking into these issues. We had the local news do a story on our food pantry which helped with the donations for a while. After, they sympathized with us but most of the action stopped. I think if we want this to stop from occurring, we need to take matters into our own hands and start changing things for the better. (Participant, 14)

> Faculty and staff are concerned about student wellbeing. We understand that COVID changed the ways students consume and purchase food. Now they're thinking about their families and taking care of their jobs and the stress of graduating college with no job prospects. Before, they could be matched with an employer and an on-campus interview which we would arrange for them; but because of budget cuts and mass layoffs, they no longer have those opportunities. They are stretching every possible opportunity and working harder than previous cohorts to get those opportunities. They're navigating a world that no one has had to prepare for in the past. (Participant, 9)

Students also expressed the shift in mindset as it relates to food insecurity and the perception of career prospects after college:

> I am working with my career advisor all the time. Do I have a job lined up after I graduate? No. But I need to decide how I'm going to deal with the unknown. I know that I have to actively seek out these opportunities and try to find things. My friends have these same concerns. We had to skip a few meals and carpool together just to be able to afford gas and clothes for interviews. I can get an amazing job after I graduate, I hope but it's coming at a great cost to me now. I just hope it will all be worth it. (Participant, 22)

Before COVID, the school would loan you a suit and padfolio to go to these interviews and there was like a bus pass to take you to the city. After COVID, they stopped the rental program. To get a professional outfit for all my interviews would cost me half my paycheck and that's not including my groceries and rent money. (Participant, 10)

I have missed out on so many events and experiences. COVID helped with virtual interviews, which was great and kept me in the market competitively for about a year. To be competitive, you must go to networking events and show up to these dinners and social things. If you don't, you're out of the running. (Participant, 3)

I often attend class tired and hungry. I am responsible and I know my food insecurity does not determine my success later on. But my mother always reminds me of women's right to work and the limited opportunities she had in her day. I never take my education for granted, and I'm not certain how I can complete my degree without constant shame and struggle. (Participant, 18)

I haven't been to many networking events since COVID, but the food pantry at my campus hosts community gardening events where I've met a lot of great people. My first internship was made through one of these events. It's been a great experience so far. I know it's not easy because I have a disability and if it wasn't for these support systems, I would not be able to succeed. I know many people like me who are on their own and are barely getting by. I feel lucky in many ways, but it shouldn't have to be this way. (Participant, 13)

I will see people cut corners in their schoolwork and use food insecurity as an excuse for late work. Grade inflation is a very serious problem that we all know about since COVID. After my second-year teaching, I realized that part of the problem was the major lack of support and food insecurity in every department. I saw students sleeping in their cars, coming to some of our events just for the food, and I'm thinking 'how did this happen on our watch?' (Participant, 16)

We had the pantry on campus open during the pandemic, we started to realize that the students were less likely to feel stigmatized on getting free food and coming to the pantry frequently, and it really shouldn't have to be this way. No one should have to be too hungry while they're finishing school, and no one should have to feel shame to get their basic needs met. It shouldn't have had to take a pandemic to make this known to us. (Participant, 9)

Another participant shared similar feedback after reporting that their campus conducted a post-covid needs assessment to their students in the educational leadership department:

> We wanted to put many things out there. COVID gave us this great experiment period. We did a lot, and I think we got the point across. People are hungry, they need to get fed. They want a decent education and a way to make a living with integrity and without struggling every day. COVID made us pay attention. Do we have a long way to go? Yes. Are we doing things now that we wouldn't think of doing 3 years ago? You bet. (Participant, 7)

A PhD student in a sociology program also noted

> We have had a different picture in mind about space and place in our society. Our lab is investigating the issues of global poverty and change from a lens of equity. Yet, we know students who skip meals do whatever it takes to get out of the cycle. But how do they complete schoolwork and find a job when focusing on everything else? If the instructor is proactive in showing mentorship early on. When there's the conversation of social capital of someone introducing you to the resources you need, how do you sustain that? If you don't have the social capital to get connected to your basic needs, there's that conversation, and what do you need to offer place-based support at the institutional level? Is there even an infrastructure in place to support that need? (Participant, 18)
>
> I currently work in a rural area. I work whenever they need me. The farm teaches me life skills that I can use in any job. It also taught me the value of food and knowing where my food comes from. (Participant, 17)

A lack of food items on and around college campuses contributes to the growth of food pantries that do not improve student populations' dietary needs, furthering the persistence of food insecurity among students. Due to the lack of food options in several campus neighborhoods, a campus food pantry is likely to be the most effective service to implement in a college or university. Food system should be structurally, economically, and geographically diverse; and most importantly, identity inclusion within food systems, within communities, and throughout society in education and beyond should be responsive to the collective identities represented in a population. It should be proactive and introspective, instead of reactive. Identity inclusion involves the awareness of knowing what student success looks like and how that develops one's character and professional outlook and cultural awareness.

If there is tuition rising, and increasing rents, and the cost of food expanding, the value of a degree in comparison to the investment on gainful employment does not feel as promising. The pressure of going to college to be successful and getting a decent job that offers a level of security and stability appears to be an illusion. To make seats available near their community, to open access and expand opportunities and open the doors wider usually implies that there is another way to ensure not everyone gets through. Even if they work hard, exhibit good character, and do everything properly and professionally, it feels like a perpetual audition where each succeeding generation is asked to do more for less.

Hunger becomes consuming of one's thoughts and the physical, emotional, and mental stress becomes traumatic, where there can no longer be brought in a set of diverse viewpoints and opportunities to allow multiple demographics have longevity to the educational output and the resilience to respond effectively to life's challenges. This expands the perpetual audition where students and society face of putting more effort into their concerns with no results or return. The quality of food at campus food pantries and the campus food environment will continue to affect students who encounter barriers to food access. Immediate changes in the retail food environment in neighborhoods and the convergence of public health concerns can also assist educational policymakers and researchers identify immediate risks associated with academic disparities.

Dietary diversity as well as rural dietary intake provide multiple ethical perspectives such as the principle of welfare (Brom, 2000), justice (Mepham, 2000), and virtue (Bobier, 2021) approaches. From the production of food to the distribution and consumption of food, the ethical implications have a positive impact on educational and societal advantages toward sustainability and ethical awareness for the common good. The principles of virtue are often perceived favorable as it relates to moral concerns and positive ways to make impactful changes. Individuals appreciate being given leadership opportunities and want to have an impact in their communities early on.

The way in which grocers and buyers effectively engage local suppliers in the food system also impacts ethical awareness which involves attending local events, visiting farms or production facilities, and establishing regular communication channels. By clearly communicating expectations, quality standards, and ordering processes to ensure mutual understanding and efficiency universities can participate in or sponsor local food-related events, such as farmers' markets, food festivals, or community workshops, to show support for local suppliers and connect directly with them. Buyers should also simplify ordering processes and logistics to make it easier for local suppliers to supply their products as it makes its way to the campus food pantry and greater institutional community.

Colleges and universities can feature local products in the pantry to increase visibility and patron awareness. They can also work with the local suppliers to improve quality, packaging, or presentation as it relates to the promotion and marketing of Halal, Jain, Kosher, Vegan, and other dietary foods. This feedback loop can enhance product offerings, inclusivity, and increase satisfaction. To make this process effective, suppliers must be willing to adapt product offerings based on seasonal availability and consumer demand and work collaboratively to educate staff about the benefits of local sourcing and encourage them to communicate these benefits to students. Provide information about local suppliers and their products to enhance customer engagement and loyalty.

By actively engaging local suppliers in these ways, grocers and buyers can foster a strong local food system, support the community, and differentiate themselves in the marketplace through unique and high-quality offerings. Consumers can engage local suppliers in the food system in several meaningful ways such as purchase directly from local farmers' markets, co-ops, or stores that prioritize local products, sign up for university student success programs where they receive regular deliveries of seasonal produce directly from local farms. Faculty can participate in farm tours as it relates to their course curricula and content and host field trips organized by local farms to understand their practices and products better. This fosters community and helps those who wish to stay in the community after graduation (Hossfeld, 2024).

Dietary diversity is crucial for ensuring optimal nutrition and overall health, particularly in rural areas where access to a variety of foods can be limited. In many rural communities, traditional diets often consist of staple crops, which can lead to insufficient intake of essential nutrients. While local agriculture may support the consumption of fresh produce, socioeconomic factors frequently hinder access to diverse food options. Limited income, transportation challenges, and lack of grocery stores contribute to a reliance on processed or less nutritious foods, impacting the health and well-being of rural populations. The ethical implications of this situation are significant. First, the lack of dietary diversity raises questions about food justice and equity. Vulnerable populations, often facing economic hardships, are disproportionately affected by food insecurity and malnutrition. These disparities call for a commitment to ethical food systems that prioritize access to diverse, nutritious foods for all, regardless of geographic location. Moreover, promoting dietary diversity in rural areas can enhance local economies through the support of sustainable agriculture and food systems. Encouraging the cultivation of a variety of crops not only improves nutritional outcomes but also fosters biodiversity, which is essential for environmental resilience. Ethical considerations should also encompass the sustainability of food production practices, ensuring that local farmers can thrive while protecting the ecosystem. Additionally, culturally appropriate foods play a critical role in dietary diversity. Recognizing and respecting local food traditions can enhance the acceptance of diverse diets, making it easier for communities to adopt healthier eating patterns. This approach emphasizes the importance of collaboration among stakeholders—farmers, policymakers, and public health officials—to develop strategies that promote both dietary diversity and ethical food systems in rural contexts. Ultimately, addressing these challenges can lead to healthier communities and a more equitable food landscape. By actively engaging with local suppliers, consumers can strengthen their community's food system and promote sustainability.

References

Bobier, C. A. (2021). What would the virtuous person eat? The case for virtuous omnivorism. *Journal of Agricultural and Environmental Ethics, 34*(3), 19. https://doi.org/10.1007/s10806-021-09860-5

Brom, F. W. (2000). Food, consumer concerns, and trust: Food ethics for a globalizing market. *Journal of Agricultural and Environmental Ethics, 12*, 127–139. https://doi.org/10.1023/A:1009586529518

Hossfeld, C. M. L. (2024). "I ain't got no extra money and I'm used to having extra money": The capacity for low-income Pickens county, SC households to manage food security during the COVID-19 pandemic. *Humanity & Society, 48*(1), 3–24. https://doi.org/10.1177/01605976231195301

Mepham, T. B. (2000). The role of food ethics in food policy. *Proceedings of the Nutrition Society, 59*(4), 609–618. https://doi.org/10.1017/S0029665100000860

CHAPTER 6

Nutritional Access: Bridging the Urban Rural Divide in Academic Progress

Abstract Nutritional access to foods is a critical factor influencing academic progress among students. Adequate nutrition is essential for cognitive function, concentration, and overall health, yet many students face barriers to accessing nutritious food. Food insecurity, often exacerbated by socioeconomic challenges, can lead to insufficient dietary intake, impacting academic performance and mental health. Students with reliable access to healthy foods are more likely to engage in their studies, attend classes regularly, and achieve better grades. Conversely, those experiencing food insecurity may struggle with fatigue, stress, and lack of focus, hindering their educational outcomes. Addressing nutritional access is therefore imperative for fostering academic success, promoting holistic student development, and ensuring equitable opportunities for all learners.

Keywords Educational performance • Stigma • Transportation • Fundraising • Access • Affordability

Food insecurity may also be more prevalent among students of color and first-generation college students (Dubick et al., 2016; Wood & Harris, 2018). Overall, food insecurity can vary among students by race, age, and employment status. According to some studies (Chaparro et al., 2009; Patton-López et al., 2014), food insecurity among students in colleges

S. Sharififard, *Community, Food Insecurity, and a Global Perspective on Campus Food Pantries*,
https://doi.org/10.1007/978-3-031-78223-7_6

and universities correlates hunger with student race and income. In a 2016 Hunger on Campus report, 57% of African-American/Black students faced food insecurity, compared to 40% of non-Hispanic White students. The report also stated that most students who experience food insecurity work, receive financial aid, and enroll in a campus meal plan. Thirty-two percent of students experiencing food insecurity reported that food insecurity affected their educational performance, and 55% of these students did not have the funds to purchase required course textbooks. White and African-American/Black students enrolled in 2-year schools are significantly more likely to experience food insecurity. White and African-American/Black students enrolled in 4-year schools are significantly less likely to experience food insecurity.

Also, at the start of their first year in college, African-American/Black students' food choices are more likely to be driven by day-to-day social activities than non-Hispanic White, multiracial, and Hispanic/Latino students, such as campus activities, fundraisers, volunteerism, and social engagement. The influence of a significant other was also more critical for White and multiracial students' food choices than African-American/Black students' (Vilaro et al., 2018). African-American/Black students were more likely to be the first in their families to graduate from college. They felt that their friends, families, peers, and society would hold a more profound respect for them if they earned a degree. Therefore, the motivation to earn a degree outweighed their hunger concerns. If the students in that population were experiencing food insecurity, they were also willing to make sacrifices to obtain a good education.

Racial demographics have also been found to contribute to concerns in higher education settings among populations experiencing food insecurity. Students from low-income and racial/ethnic minority families are at an increased risk of experiencing food insecurity during college enrollment (Payne-Sturges et al., 2018). In a survey conducted by the City University of New York, about 39% of students reported that they had experienced either one of several forms of food insecurity. The researchers also found that African-American/Black and Hispanic/Latino students were more likely to experience food insecurity than Asian and White students at the same university. Further, students who supported themselves financially were more likely to report food insecurity; and students working more than 20 hours per week had a higher rate of food insecurity than those who were unemployed.

Twenty-four percent of students reported experiencing both food and housing insecurity. Students over the age of 21 were more likely to have both housing and food insecurity than students who were not receiving work-study support. Forty-five percent of students reported concerns about not having enough money to purchase food. Twenty-two percent of students reported that due to a lack of funds, they were often hungry. The researchers further found that students who also experience mental health problems are more likely to have experiences with food insecurity (Freudenberg et al., 2013).

Between March and May 2016, four campus-based organizations, including the College and University Food Bank Alliance (which was acquired by Swipe Out Hunger in 2021), the National Student Campaign Against Hunger and Homelessness (Students Against Hunger), the Student Government Resource Center, and the Student Public Interest Research Groups, surveyed students about food insecurity. Students were recruited from 12 states attending community colleges and four-year colleges and universities. Fifty-seven percent of African-American/Black students in the sample reported experiences with food insecurity. Forty percent of White students, and 56% of all the surveyed first-generation college students, also reported experiencing food insecurity.

Further, community colleges and the 4-year colleges and universities in this study may have been in rural, urban, and suburban areas. It is important to distinguish among each community to determine the levels of food insecurity in each neighborhood. Mobility and migration patterns are also likely to influence similar studies. First-generation students and their migration patterns before entering college, as well as their parents' and relatives' past migration, is also important to differentiate between neighborhoods and food supply.

A study examining the relationship between race and food insecurity among college students revealed that 61% of African-American/Black students were experiencing food insecurity compared to 32% of White students (Fetter & Gilboy, 2018). Students who identified as Hispanic/Latino or multiracial were found to experience more food insecurity (Duran & Nunez, 2020). Female students in the sample were also more likely to experience food insecurity, but there were no relationships between females experiencing food insecurity to males.

Students living with parents or relatives may have a consistent source of food available, which helps with their food insecurity status. In contrast, those living with roommates, a significant other, or solo may have difficulty with sharing multiple responsibilities, financial management, and budgeting. Such expenses as rent and utilities may be fully covered when a student lives with parents or relatives. In other living arrangements, the student might be responsible for all their expenses with no assistance. Further, if a student was also responsible for childcare, they were more likely to experience food insecurity because of the costs of taking care of their child.

The decisions a student makes to attend a school reflect the opportunities they will have in those communities. Demographic shifts and poverty contribute to the success or failure a student may encounter; these developments are likely to intensify as a result of the pandemic through decades to come, and therefore create complexities around improving academic outcomes. Racial inequalities act as structural determinants shaping access to food, and in many cases, to food that is healthy, appealing, and affordable (Maguire, 2016). According to Helling and Sawicki (2010), when African-American/Black neighborhoods are primarily poor, the insufficient availability of retail stores affects the lack of presence of grocers and businesses offering quality produce in these neighborhoods. The researchers also indicate that such disadvantaged neighborhoods provide less accessibility to nutritious foods with smaller grocers, and residents have a longer commute to larger supermarkets with fresh produce than predominately White neighborhoods. Neighborhoods with higher proportions of African-American/Black residents have more fast-food restaurants and fewer supermarkets than neighborhoods with higher proportions of White residents (Hilmers et al., 2012).

Non-fast food restaurants are more prevalent in racially mixed and White neighborhoods than in African-American/Black or Hispanic neighborhoods (Dimitri & Rogus, 2014), which influences the diet and food quality of individuals residing in those neighborhoods. Other researchers have found no difference in diet quality because of living closer to a supermarket (Larson et al., 2009; Raja et al., 2008; Moore & Diez Roux, 2006; Morland et al., 2002) but have considered food environment measures for neighborhood characteristics. Institutes of higher education may significantly differ in the types of food items they can offer on campus. Other missing analyses included the prevalence of public transportation, which would affect safety and neighborhood behaviors practiced by first-time students in large cities who are newly acquainted with the public transit system and the neighborhood with a traditional market presence.

Further, low-income, minority, and immigrant communities that are disparately affected by food production and restrictions particular to their diets and religious accordance (Neff et al., 2009) may also encounter challenges with the commute to an accessible grocer. The zip codes of such communities and their proximity to a variety of foods are of critical importance to the types of barriers that may exist in those sectors, as well as their relationships to diet quality and neighborhood safety (Bader et al., 2010).

Kamarulzaman et al. (2016) implicate that only several large urban cities, and only a few major university towns in the U.S., have adequate food choices for individuals following a Halal diet. Students who study at schools within areas lacking the foods needed to sustain such a diet may have limited to no access to such foods. Notably, international students who follow similar diets may also encounter difficulties with finding cultural foods within proximity to their school; they may or may not be willing to commute to a place where such options are available.

As international students assimilate to their host country's customs and culture, they are also more likely to change their practices as well as the nutritional content of their food choices. In a 2015 study, researchers found that most international students in the United States drank caffeinated beverages and skipped meals to compensate for their food insecurity. The students had indicated that traditional food items are different in the United States from their home country in terms of price and quality (Alakaam et al., 2015).

Further, traditional stores are likely to be in larger cities. Students who reside in these cities often rely on public transportation, which may be limited to some geographic areas and university towns. These nonacademic barriers also affect campuses in rural areas and often have similar issues and barriers that other campuses in differing geographic areas have in terms of food provision (Mirabitur et al., 2016; Waters-Bailey et al., 2019). In such cases, students would have to rely on resources available on campus. However, students specifically practicing a Halal diet in the United States had limited access to such items on campus and in the surrounding neighborhood, which led them to consume fewer meats in their diets.

Ethnic markets and similar specialty stores are likely to be more important to specific communities and populations. A group of people who are hesitant and unwilling to make purchases at large traditional markets might feel excluded or uncomfortable because their preferred items are superseded by unfamiliar food choices and ingredients (Joassart-Marcelli et al., 2017). Such food practices may also be met with perceived stigmas within a community that makes ethnic groups feel uncomfortable and unsafe in the environments and neighborhoods in which they reside.

In the long-term, these perceptions may affect where an individual decides to attend college. On the contrary, students affected by food insecurity might not be willing to purchase items from an ethnic or specialty store located near his/her campus because of perceived stigma, uncomfortable emotions, and lack of familiarity with the food items and ingredients relevant to their individual needs. It is important to consider that no single market or food store would necessarily carry the item(s) appropriate to one's needs. It is also as important to know how likely a student is willing to continue to travel the same or different distances to get the food required for their diet. From a cultural perspective, acquiring and maintaining the foods a student needs because of what they are used to consuming can affect their food preferences, choices, and subsequent lifestyle changes.

Understanding the depth of food insecurity and the outcomes of academic success can help researchers understand how students cope with experiences of food insecurity. A survey (McArthur et al., 2017) of 6000 undergraduate and graduate students asked students to indicate how often they had coped with their food insecurity status. Undergraduate freshmen were not included in the sample. The researchers sought to understand if students had been affected by difficulties with food security and academic progress with factors such as understanding course concepts, graduating on time, and students' attention span during class when experiencing food insecurity. The researchers calculated the participants' BMI from self-reported height and weight figures. Unlike other studies, this study's survey validity was determined by nutrition professors at the university who had also had professional experience with food security. The researchers indicated that about 46% of students were experiencing food insecurity (McArthur et al., 2017) but did not report if a cohort of students—i.e., undergraduates or graduate students—were found to have experienced more food insecurity by the self-reported responses.

Approximately 60% of students in the sample were either undergraduate juniors or seniors. Ninety percent of students were reported to have experienced food insecurity. Of the students experiencing food insecurity, 66% were female, and nearly 90% were classified as full-time students. Seventy-five percent of the students had personal monthly incomes totaling less than $500; 70% had received financial aid; and an overwhelming

75% of students had not purchased a university meal plan (McArthur et al., 2017). The students who were reported to have declined purchase of a meal plan and who reported experiencing food insecurity while in school may have had trouble affording such a meal plan and might have sought other alternatives. Additionally, the students reported having experienced food insecurity at the time of the study indicated various types of support that they perceived would help increase their access to food: securing a job, receiving additional financial aid, budgeting, eating healthily, and shopping for food (McArthur et al., 2017).

The researchers found strong relationships when comparing students experiencing food insecurity with academic variables, where GPA was negatively correlated, which suggested that students experiencing food insecurity tend to be met with less academic success. A lower percentage of students experiencing food insecurity perceived their academic performance, their understanding of course concepts, and their absenteeism rate as excellent or good as compared to their food secure counterparts (McArthur et al., 2017).

In a study consisting of a predominantly White population at a land-grant university in New Hampshire, 47.6% of students receiving financial aid reported having an unlimited meal plan, 29.4% reported having a plan with a set number of meals, and 23% reported having no meal plan (Davidson & Morrell, 2018). Sixteen percent of those respondents had qualified for free or reduced-price school breakfast/lunch when enrolled in elementary, middle, and high school, or at some point in their primary education.

Approximately 4.3% of students reported they had or had currently been receiving food stamp benefits, and 42.4% reported not knowing if they qualified for the benefits. Only 7.2% of students reported using the services of a food pantry or other food assistance program. The study did not find a relationship between food insecurity and GPA but suggested that students with a GPA higher than 3.0 had the lowest food insecurity. In contrast, students with GPAs between 2.0–2.9 had the highest rates of food insecurity. However, students with less than a 2.0 GPA did not experience high rates of food insecurity, which further suggests that the students may have been experiencing other academic and personal issues that were not relevant to their food insecurity status. Students in the highest GPA range, classified as 3.5–4.0, were compared to students in a lower GPA range, between 2.0–2.4, there was a significant relationship between food insecurity and student GPA. Students experiencing food insecurity were significantly less likely to fall into the higher GPA category. Students who did not experience food insecurity were more likely to see positive effects in their GPA.

In another study examining the relationship between food insecurity and academic achievement, 38% of students experienced food insecurity at the end of the school year (van Woerden et al., 2018). The GPA of students experiencing food insecurity was lower than the GPA of students not shown to have experienced food insecurity. The sample of the undergraduate freshman students included 70% female and 50% non-Hispanic White. Overall, there was no relationship between food insecurity and retention. Some of the limitations of this study were that the number of maximum credit hours per semester for undergraduate students was not constant across students. Further, by using a cumulative university GPA, the effect of food insecurity by the end of the school year varies.

As found in Fetter and Gilboy (2018), students who reported experiencing food insecurity had a higher likelihood of either dropping out of school or reducing their academic units. Students who worked full time were also more likely to reduce their academic units than students who were not working full time. However, in some other situations and educational environments, experiencing any degree of food insecurity does not inform the importance of students experiencing overall academic success (Meza et al., 2019). At a large public research and land-grant university in northern California, more than 8000 participating students were surveyed about food insecurity and their GPA status (Martinez et al., 2018a).

In the distributed questionnaire, students were asked to rate their level of stress, and indicate if they had ever felt any of the following emotions as a current student:

1. Hopelessness
2. Overwhelmed
3. Exhausted
4. Very lonely
5. Very sad
6. Overwhelming anxiety
7. Overwhelming anger
8. Depression (Martinez et al., 2018b)

Overall, 40% of students reported experiencing food insecurity. Ninety-three percent of those students had significantly higher proportions of poor mental health compared to their peers who had not experienced food insecurity. Intergenerational inequity also impacts mental health outcomes because of food insecurity beyond higher education (Loofbourrow & Scherr, 2023). Being resilient and successful in higher education becomes less common and even burdensome as so much of determining what might be appropriate overtime is determined. The distribution can be aligned with the organic, social perspective where students and patrons appear to be more responsible for selecting foods that are not high in fats and sodium. Providing food sometimes means finding creative solutions like home delivery, and greater selection to ensure needs are met in a way that engages students with progress, not survival. In class discussions, this becomes normalized and at a micro-level more activities are created in a collaborative space. Systematic review of the effect of these policies reported that most studies showed improvements in students' diets as access to healthy foods and beverages in the school environment increased. There is no published literature on higher education nutrition policies; however, a review of nutrition interventions in college settings found positive associations with dietary quality (Rickrode-Fernandez et al., 2021).

Fifteen percent of the students reported experiencing tremendous stress. 28–55% said they had felt very sad, very lonely, hopeless, overwhelming anxiety, overwhelming anger, and depression, making it difficult to function. Between 76–82% felt overwhelmed and exhausted as a result of their food insecurity. On average, students in the sample were 23 years old, female, and either White, Asian, or Hispanic/Latino. Most of the students were reported to have a GPA in the 95th and 80th percentiles, which suggests that students affected by food insecurity would not let their individual issues, such as food insecurity, interfere with their academic success (Wolfson et al., 2024).

Other researchers used a non-random sample of students enrolled full-time in a four-year university. Nineteen percent of students were classified as very low food secure and 16% as low food secure, but 64% of students were classified as food secure (Wooten et al., 2019). The students self-reported an average GPA of 3.41, 81% of participants identified as White, and 70% identified as female. Students with a self-reported GPA between the ranges of 3.8 4.0 were less likely to experience food insecurity. Students with a lower self-reported GPA were significantly more likely to experience food insecurity than students with the highest self-reported GPA, suggesting that academic success is more significant than food insecurity in college. As the GPA scores were all self-reported, there may have been data in the study, where students may have exaggerated or inflated their responses. Further, they may not have considered a cumulative GPA toward degree completion.

A student who reported receiving loans that required repayment was more likely to be experiencing food insecurity. A student who received scholarships that did not require repayment was more likely to be food secure (Wooten et al., 2019). Students who had been employed either part- or full-time were less likely to experience food insecurity than students who were not employed. Students who used personal savings as an income source had an increased likelihood of food insecurity. The student enrollment status was also not described in this study. When controlling for race, age, gender, and ethnicity, students who were experiencing food insecurity before entering college were more likely to experience food insecurity during college. Based on these findings, having financial independence cannot be the sole predictor of food insecurity at colleges and universities. Financial independence also cannot determine whether a student had had past experiences with food insecurity before enrollment. The study did not confirm if scholarships covered tuition and ailment needs, or if students were eligible for renewing different types of scholarship monies deposited directly to the student's account rather than the institution. There was no significance between students who participated in a campus meal plan and those who did not (Wooten et al., 2019).

The same students in the study increased spending on non-essential items. They also had a significantly lower likelihood of food insecurity. However, when students increased spending on essential items, they were more likely to have had an increase in food insecurity. Students with food insecurity may have had more disposable income or had the support from parents or relatives to pay for their essential needs. However, the study does not identify essential needs and items and does not indicate the value or utility of such necessities across students.

For example, a student enrolled in one program might be required to purchase different books and supplies for their course requirements than a student who has other curricular requirements. Non-essential items could also be referred to as optional course requirements, which students determine for their own academic needs. The expenses incurred in this sample may be related to other studies whereby employed students distributed much of their income to family members and others who were overlooked. Further, the results did not explain if some of the employed students were part of any federal work-study programs.

Current living conditions and previous household characteristics were prevalent before college entry and are also essential to the analysis and understanding of food security cycles in a period. Since the study used a self-reported survey method to collect data, biases could have been presented in the results. The researchers also confirmed that the response rate is low compared to the university population. They also reported a higher percentage of female respondents in the sample, which is more than the overall university population (Wooten et al., 2019). The use of nonrandom sampling obtained in this study may not be representative of the university population, which may have resulted in selection bias.

The recognition of student awareness about food insecurity and environmental issues around campus also explains how involvement in specific campus organizations and programs helps raise awareness about the university setting. Additionally, the students can understand the role of stakeholders and university leadership. These factors help support campus needs and emerging areas of concern that affect student academic success.

Students from two undergraduate English courses, and a group of students in service-learning programs, in a public research university in Ohio, surveyed their campus about student food insecurity (Twill et al., 2016). With more than 3700 student respondents, 49% stated that they had experienced some degree of food insecurity (Twill et al., 2016). As an overall response to the survey's results, the student government association proceeded to ask school administrators to provide space and resources for a campus food pantry to alleviate the issues of food insecurity and hunger. During the first two years of pantry operation from 2011–2013, there were a reported 870 visits to the pantry, of which half were repeated visits. African-American/Black students made up 50% of visitors, followed by about 26% of White students. Pantry users were predominantly female (69.9%), enrolled full-time (90%), lived off-campus (53.3%), or were reported to have lived with children or a spouse/partner (17.6%).

It was further reported that the largest college at the university, the liberal arts college, had the most pantry users. First-year students were found to have used the pantry somewhat more than their upperclassmen peers have, and 19% of pantry usage was made up of graduate students. Other statistics about the patrons included that 45% were unemployed and could not afford food and had difficulty with budgeting. Pantry users who were employed also reported problems with being able to afford food and manage a personal budget. Twenty-six percent of patrons had a scholarship, though it was not confirmed what type of aid was covered or if the funds were given directly to the students. Sixty-two percent of student respondents used loans, and some reported receiving financial assistance from their families (Twill et al., 2016). Other pantry users also lived with roommates, parents/relatives, or alone, indicating that the pantry served a variety of students who needed food assistance. Students are also responsible for staffing the pantry to ensure their peers are comfortable with using the service.

Other student-run initiatives include *Swipe out Hunger*, a program launched in response to helping students and dining halls implement systems whereby students could donate their extra meal points to other students who need meals. As another temporary solution, some students have developed mobile applications or websites to enable contributing students to use social media and related technology capabilities to donate or share meals with other students. For example, in 2015, students at Columbia University in New York created a mobile application called *Swipes* to accomplish the same goal of sharing meals and delivering access to information about campus resources that offer help with food assistance (Chiu, 2015). However, it was reported that after a year of launching, the application had not improved the consequences of food insecurity, and had non-functional features and systemic bugs in the app.

At Tufts University, a private research university in Massachusetts, students established the *Swipe it Forward* program where students would learn to feel more comfortable with communicating their needs and food insecurity concerns. Students also can acquire food through donated meals from other university students who have not used their meal plan points during an academic term. At Northeastern University, another private research university in Massachusetts, students can opt-in to receive texts to get edible leftover food at the campus that would otherwise be discarded (Kornwitz, 2017).

One barrier to this opportunity is that some students may not have access to mobile devices to receive text messages, or to other electronic devices. Also, the cellular carrier may have different fees for receiving text messages that might affect how students hear about the program. Other problems might consist of downloading a compatible mobile application, using all the apps' features, and having the technological capability for messages continuing to be identified as spam under unreliable networks. Using a university computer may violate some of the school's technology usage policies. It may expose risk in tracking individuals in need of food assistance and students with chronic food insecurity. Such programs may result in a series of unintended consequences for students who are concerned about revealing their identities or sharing their personal experiences about food insecurity.

A study of students at a large, land grant university in Appalachia found that about 36% of 456 respondents were experiencing food insecurity (McArthur et al., 2018). The study looked at families and campus food insecurity among freshmen students. An online questionnaire contained socio-demographic items that measured food insecurity, academic progress, coping strategies for accessing food, and social support. Participants primarily reported housing and food insecurity. About 42.5% of students who experienced food insecurity believed their access to food had worsened since starting college.

Students who were reported to have been food secure were also found to have had more significant academic progress and had perceived their individual eating habits since starting college as healthy/very healthy and perceived their health status as good/excellent. Other students in the study reported requesting assistance with job opportunities, affordable meal plans, financial management, and healthy eating (McArthur et al., 2018). Thirty-two freshmen students also reported housing and food insecurity.

To identify relationships between students who had experienced food insecurity in households they grew up in and students discovering food insecurity on their own during their time on campus had not been assessed. It is then suggested that student food insecurity is more likely to begin during the first year of college, so there is a need for interventions to increase food access. Both institutional and community efforts would be necessary to approach such a goal to encourage students to use campus resources, as well as community partnerships, if there is a lack of services among campus.

Another study consisting of students attending a university in central Appalachia found that food insecurity was highest during undergraduate sophomore and junior years, and lowest among graduate students. Students who lived off-campus were found to have experienced more food insecurity than students who lived on campus (Hagedorn & Olfert, 2018). Like other studies, biases and inaccurate data may have been reported as students may have felt shame and embarrassment from reporting actual and truthful personal experiences.

As a coping strategy, students may have chosen to edit some of their responses because of overwhelming affirmation toward a specific position on their health status. Further, rating individual health status based on descriptions, such as poor or fair, does not describe solutions to remedying some of the barriers persistent in overall health and wellbeing as it relates to food consumption. It also does not indicate food insecurity status. Respondents may have also interpreted food insecurity as a personal term related to individual lifestyle and dieting, rather than its actual terminology.

Campus food pantries in the U.S. have increased within the past decade. In 1993, Michigan State University (MSU) established the first college food pantry on campus, named the *MSU Student Food Bank*. In 2012, MSU collaborated with the Oregon State University Food Pantry and established the College and University Food Bank Alliance (CUFBA) to assist campuses with starting their food banks and addressing student hunger. As of 2019, the CUFBA has more than 700 registered member institutions (MSU student food bank, 2019), of which about 253 members have established operating food pantries on campus. Being a registered CUFBA member means an institution seeks information and assistance about food insecurity issues and has considered developing a food pantry on campus. There are also campus food pantries that have not registered with the CUFBA, which indicates a larger presence of food pantries at institutes of higher education. The increase in institutional CUFBA membership portrays the urgency of food-insecurity issues affecting college campuses.

While only a temporary solution, campus food pantries help students seek different options for their food insecurity needs without encountering some barriers to seeking aid elsewhere. Community food banks also help to relieve unanticipated situations and employ volunteers and staff members to manage the pantry's operations. Despite the growth, the success of food distribution services is as successful as its nutritional content of food, an understanding of the needs of patrons, and the knowledge of the prevalence of hunger and food insecurity. Since there is no preliminary requirement to travel off-campus, campus food pantries may serve as an appropriately convenient option for students to access the pantry. Food pantries also increase the likelihood of student patronage compared to community-based programs when identified as a student resource or service.

Campus pantries also reduce the risk of identifiable personal information that would potentially be exposed through mobile applications or websites promoting sharing meal plans.

According to Cady and White (2018), campus-based pantries are stocked through direct purchases and food donations. If a pantry purchases food, a partnership with a state or regional food bank can be developed whereby campuses partner with a nonprofit agency or an organization that enables the campus pantry to be part of the food bank network through the sponsorship. Food donations are also provided by local agencies, campus student groups, regional, national, and multinational businesses, university alumni, faculty, and staff.

On campuses where donations are the primary source for stocking the pantry, there are often annual events such as fundraisers, galas, and special occasions that contribute to large amounts of the food supply. However, food supplies from donations are less predictable, and some of the donated food may be expired or in an undesirable condition prohibitive for consumption. Another issue with some of the food pantries are the efforts of staff members to check the quality of the donated foods to ensure they are in nonperishable condition. Proper storage such as extending the freshness of fruits and vegetables could reduce spoilage. Instead of discarding foods prematurely, the pantry staff can use it for compost for gardening.

Donations can be unpredictable and may not always meet the needs of students. Pantries may not have the technology or infrastructure to efficiently manage inventory, track client data, or provide online services. Due to sickness and liability issues, some food cannot be donated, and dining halls are forced to discard food in good condition. Campus pantry personnel and campus officials often collaborate with academic departments to raise awareness about the issues and problems facing students who are dealing with food insecurity and promote programs that support these students through other campus facilities. Empty bowls fundraisers are also typical supporters of campus food pantries where students enrolled in art and culinary programs donate supplies, such as ceramic bowls, and host events where admission fees are awarded to the campus food pantry (Price et al., 2019).

Yet, colleges and universities have a financial incentive to maintain. By charging a fair amount of expenses for a meal plan or partnership for a meal delivery service can be seen as an investment in their education and might pay off. However, the building investment of the physical space on campuses might not always cut into profit margins as the world observed during the pandemic where students were primarily taught in fully online or hybrid modalities, which became a real estate issue. As a result, more food pantries closed, campuses shut down, and more pantries began reducing their inventory.

If a food service provider such as Sodexo agrees to support and endorse a wellness center on campus, here is a financially valuable contract that can make a return for the institution that other projects cannot. In Japan, there is great production and consumer interest, especially in beverages where product assortment is much broader than in the U.S. and Europe. In Europe, the interest on the part of producers and consumers is mainly directed at gastrointestinal health. A sushi bar on campus might have a financially valuable contract, seen as a sustainable source of revenue instead of a relevant student need that is obligated to be met as part of student success responsibilities. On-campus foodservice generates over $18B in sales revenue. The average college and university charges about $16 per day, and up to $38 a day, for a three-meal-a-day dining contract, compared to the less than $11 a day a single person spends (on average) for food. Students' basic needs have been recognized as a potential revenue stream that can help with the budget cuts due to COVID-19. Universities often opt for contracts with large food service corporations that promise financial returns to the university. This exchange promises financial returns to universities who will see an increase in their market.

Campus pantries are in a variety of spaces around campuses. Some common locations include student affairs offices, counseling centers, health centers, academic divisions, and administrative divisions or units. The social environment contributes more to health disparities than the physical environment, such as buildings and the absence of sidewalks (Keith et al., 2015). It is essential to consider that even if there is adequate availability of food within short proximity to a college campus, a lack of transportation and a lack of safe sidewalks will affect how students acquire their meals. Students who cannot commute to a grocer around campus neighborhoods would see no immediate advantages to their community's food environment. In such cases, a food pantry would be a better or more convenient solution, assuming that the campuses and the food pantries are in areas where all students have safe and accessible transportation modes. However, it is essential to consider how such conditions and spaces around campuses affect students with disabilities as well as students with barriers to traveling from place to place.

Pantry staff members often have several responsibilities in addition to leading their current positions such as daily pantry operation, familiarity with rules, procedures, and protocols of the pantry's management. The staff may sometimes be new to managing pantry facilities or might have little to no experience and knowledge about food insecurity, student resources, and the availability of community resources (Price et al., 2019), which may make the pantry's services less effective. Many food banks require member pantries to serve the public as part of the food bank network, but many universities may not be willing to open their campus food pantry services to the public (Cady & White, 2018).

A preliminary survey of students from the student government, the college honors program, an undergraduate level English course, and students in the social work major, at a 4-year public research university, was distributed to assess whether a food pantry was needed on campus. The survey served to determine if a campus food pantry would help students with food insecurity issues (Reppond et al., 2018). Social work and undergraduate-level English class students were recruited for this study because the researchers believed that the individuals enrolled in those programs would have more familiarity with food insecurity issues than their peers. They also believed students from those programs and classes would be driven by some of the related problems on campus.

A limitation of this study is that classes were randomly selected. Through the introduction of the course instructors who were willing to allow the researchers to distribute the surveys as they perceive relevant to their course curriculum, which could incur selection. Having an established and professional working relationship with the researchers or having substantive knowledge about food insecurity and student wellness could be beneficial to the study. Still, the participants would likely have had more functional experience and expertise in the subject than their colleagues from other campus programs. Another limitation is that the surveys may contain insufficient or misinterpreted responses because they feared having their answers seen by peers, which would likely affect their mental health.

The study did not link data to individual student academic performance or GPA. GPA was an accurate assessment of the changes in academic achievement for students, such as time to complete a degree, class participation, and frequent absenteeism, because of food insecurity challenges. Nearly 150 students responded, of which 48% reported experiencing food insecurity, whereas 64% of respondents knew of a friend or classmate who had experienced food insecurity. Ninety percent of respondents either agreed or strongly agreed that the campus helped support students experiencing food insecurity (Reppond et al., 2018).

When asked about the patronage of a pantry on campus, 60% either agreed or strongly agreed that they would use the pantry if necessary, and 94% indicated that they would refer others to the pantry (Reppond et al., 2018). While it was agreed that a campus food pantry could successfully operate because many students were much in need of such a service, the school preferred distributing food to students who did not need food assistance. The school did not want to risk probing questions to a student who needed food assistance because they believed the process would result in the student being turned away from receiving food over the policies and matters surrounding their circumstances.

Researchers at another public university in Florida sought to examine the negative social stigmas that may prevent students from using the campus food pantry. Two hundred sixty-three undergraduate students were surveyed in an introductory sociology course (Loftin, 2013). The survey responses revealed that approximately 24% of students have had to skip meals because they could not afford food, and about 14% needed food assistance. The awareness of the campus food pantry was high. About 75% of students reported knowing about the campus food pantry, and 63% knew where the pantry is located. Still, each segment of students generally felt embarrassment and shame in accessing its services.

The same researchers also distributed 51 different surveys to students who would consistently use the pantry for food assistance. The survey had also included questions about years in college, ranging from first-year undergraduate to fourth or more-year senior. Gender and age were used to assess whether they influence perceived stigmas. The study also sought to consider the students' living arrangements and their financial history, as well as their families' socioeconomic status.

The participants were asked to indicate their difficulty in accessing the pantry on a 5-point Likert scale ranging from "very difficult" to "not difficult at all." Participants were also able to explain any factors that may have prevented them from getting assistance from the pantry, how they became aware of the pantry, and if they share the information about their status with others. About 66.7% had worked while being a full-time student, with 58% of those having to have worked full-time. Further, 58.8% of the students lived off-campus, 12% lived with parents, and about 87% lived with a roommate (Loftin, 2013).

More than half of the students lived off-campus, and almost 70% indicated that the pantry location did not affect their access to the pantry. Further, about half of the students reported spending the most money on food, followed by paying for bills and gas; and more than 70% indicated receiving assistance from their families (Loftin, 2013). When asked how they had heard about the pantry, 35.4% said through friends and by word of mouth. About 17% reported having walked by the pantry while on campus. Twenty-one percent of the students indicated having family members who had received food stamps. Over 40% of the respondents were freshmen, 31% sophomores, 16% were juniors, and almost 10% were seniors (Loftin, 2013). In a mixed-methods study (Paola & Debate, 2018) observing food pantry users (n = 221) and the relationships among food insecurity, diet quality, and academic achievement, most patrons self-reported as female, White, or Hispanic/Latino; and were primarily undergraduate juniors or seniors. About 53% of respondents were employed, and 14% reported to also having a campus meal plan. Ninety percent of visitors strongly agreed that they were treated professionally by their campus food pantry staff and volunteers, and 95% rated their experience with the pantry as very good/good, which increased their likelihood of returning to the pantry whenever needed. Eighty-seven percent strongly agreed/agreed that the pantry provided them with the foods they would otherwise have to have skipped, and 95% strongly agreed/agreed that the pantry resources enabled them to allocate funds toward rent, utilities, and medical needs.

All the surveyed students indicated that the resources provided by the food pantry contributed to their academic success. Forty-four percent of patrons used the pantry once, and about 55% of pantry users used the service more than once in an academic semester. Among self-reported food respondents who were experiencing food insecurity, half were single time pantry visitors, and the other half of students were recurrent visitors (Paola & Debate, 2018).

The researchers concluded that there was no relationship between single/frequent visitors and socioeconomic characteristics. The study also did not reveal how students using other or outside services and resources benefited from the food pantry or showed any significance about the academic development and success of their studies. Recurrent visitors also did not report if they were also using other services, and if the support of the pantry is helping with their academic and personal growth. As the survey was self-reported, there may have been biases in the responses. Further, the survey did not include questions about the living situations of the pantry users, limiting the type of analysis for this study.

Another study observing food pantries and accessibility on campus found that one of the most concerning challenges in starting a food pantry is having adequate physical space on campus. Participants in the study described how some of their institutions would not permit space on campus for the pantry. An additional challenge reported related to finding space on campus was finding support for a food pantry. Being able to convey the need for and importance of a food pantry on campus was difficult to promote, and the level of urgency needed to engage various divisions, departments, and faculty and staff was lacking across the institutions (Gupton et al., 2018). In three months, the pantry recorded 827 transactions, of which 39% were repeat patrons, and had served 504 unique students.

Of the 39% who had used the pantry more than once, overall visits ranged from 2 to 17 trips. Twenty percent of the pantry users would use the pantry multiple times during the academic year. Data on student status and residency were not reported, but the researcher's analysis revealed that 62% of its patrons were graduate students, and 38% were undergraduate students. Seventy-one percent were international students, 46% were living alone, 12% were living with one other adult, and 16% were living with three or more other adults. Sixty-two percent reported their need for food assistance as "ongoing," and 19% indicated it as a "one-time" occurrence. The highest percentage of students responded that they occasionally experienced food insecurity; 15% chose rarely, 17% chose "frequently," and 7% reported they experienced food insecurity very frequently.

In terms of pantry usage patterns, results indicated that the highest number of students visited the pantry during its first week of operations. For the remaining weeks of the fall semester, the pantry served an average of 35 students per week. During winter break, the pantry opened for two days and served only 13 patrons. During the spring semester, the pantry served an average of 28 students per week through the end of the academic term. During the summer, open pantry hours were reduced to one day per week, and during that time served an average of 12 patrons per week. The first week of operations was likely to be the busiest period of service, while students were becoming acquainted with campus facilities and adjusting to new changes on campus. The pantries on campus would also be impacted at the beginning of a semester as students become acquainted on campus. It is important to identify repeat patrons of the pantry throughout the school year to determine if persistent food insecurity affects certain groups of students or other factors contributing to the types of food pantry recipients and their food security needs.

A lack of awareness of the pantry, its location(s), and hours of operation are all contributors to the success or failure of the service. The potential stigma that may prevent some students in need from seeking help on campus may also deter students from using the services offered by their campus food pantry. An internal issue that campus food pantries may have been a lack of variety and perceived or desirable foods available at the pantry, as well as a lack of cultural inclusivity present in the food choices. For example, students may have dietary, ethnic, medical, or other personal preferences and restrictions that require more selections of food options than items available at the campus food pantry.

A lack of select items and a lack of knowledge about preparing some of the foods may also affect the success and effectiveness of the pantries. The pantry location, setup, and the availability of foods and supplies affect how students are served. Ninety-two percent of current campus pantries have a space on campus in which they operate. However, the size of the area in which the pantry is located may vary and affect the types of foods and supplies that can be stocked and distributed. It may also change the number of patrons who can be present without violating any fire hazards.

References

Alakaam, A. A., Castellanos, D. C., Bodzio, J., & Harrison, L. (2015). The factors that influence dietary habits among international students in the United States. *Journal of International Students, 5*(2), 104–120. https://www.ojed.org/index.php/jis/article/view/428

Bader, M. D., Purciel, M., Yousefzadeh, P., & Neckerman, K. M. (2010). Disparities in neighborhood food environments: Implications of measurement strategies. *Economic Geography, 86*(4), 409–430. https://doi.org/10.1111/j.1944-8287.2010.01084.x

Chaparro et al., (2009). Food insecurity prevalence among college students at the University of Hawai'i at Mānoa Abstract. *Public Health Nutrition 12*(11) 2097–2103. https://doi.org/10.1017/S1368980009990735

Cady, C. L., & White, C. C. (2018). Food pantries on campus to address student hunger. *New Directions for Community Colleges, 184*, 73–82. https://doi.org/10.1002/cc.20329

Chiu, C. (2015, November 15). *Two CC students launch mobile meal-sharing app Swipes.* https://www.columbiaspectator.com/news/2015/09/20/two-cc-students-launch-mobile-meal-sharing-app-swipes/

Davidson, A. R., & Morrell, J. S. (2018). Food insecurity prevalence among university students in New Hampshire. *Journal of Hunger & Environmental Nutrition*, 1–10. https://doi.org/10.1080/19320248.2018.1512928

Dimitri, C., & Rogus, S. (2014). Food choices, food security, and food policy. *Journal of International Affairs, 67*(2), 19–31. https://www.jstor.org/stable/24461733

Dubick, J., Mathews, B., & Cady, C. (2016). *Hunger on campus: The challenge of food insecurity for college students.* College and University Food Bank Alliance. http://studentsagainsthunger.org/wp-content/uploads/2016/10/HungerOnCampus.pdf

Duran, A., & Nunez, A. M. (2020). Food and housing insecurity for Latinx/a/o college students: Advancing an intersectional research agenda. *Journal of Hispanic Higher Education*, 1–15. https://doi.org/10.1177/1538192720963579

Fetter, A., & Gilboy, M. (2018). Addressing food insecurity on college campuses: The food pantry and beyond. *Journal of the Academy of Nutrition and Dietetics, 118*(10), A160. https://doi.org/10.1016/j.jand.2018.08.126

Freudenberg, N., Manzo, L., Mongiello, L., Jones, H., Boeri, N., & Lamberson, P. (2013). Promoting the health of young adults in urban public universities: A case study from City University of New York. *Journal of American College Health, 61*(7), 422–430. https://doi.org/10.1080/07448481.2013.823972

Gupton, J. T., Trost, J. L., & Collins, K. (2018). Food pantries as a gateway for academic enhancement and basic needs support. *New Directions for Community Colleges, 2018*(184), 61–71. https://doi.org/10.1002/cc.20328

Hagedorn, R., & Olfert, M. (2018). Food insecurity and behavioral characteristics for academic success in young adults attending an Appalachian University. *Nutrients, 10*(3), 361. https://doi.org/10.3390/nu10030361

Helling, A., & Sawicki, D. S. (2010). Race and residential accessibility to shopping and services. *Housing Policy Debate, 14*(1-2), 69–101. https://doi.org/10.1080/10511482.2003.9521469

Hilmers, A., Hilmers, D. C., & Dave, J. (2012). Neighborhood disparities in access to healthy foods and their effects on environmental justice. *American Journal of Public Health, 102*(9), 1644–1654. https://doi.org/10.2105/AJPH.2012.300865

Joassart-Marcelli, P., Rossiter, J. S., & Bosco, F. J. (2017). Ethnic markets and community food security in an urban 'food desert'. *Environment and Planning A: Economy and Space, 49*(7), 1642–1663. https://doi.org/10.1177/0308518X17700394

Kamarulzaman, Y., Veeck, A., Mumuni, A. G., Luqmani, M., & Quraeshi, Z. A. (2016). Religion, markets, and digital media: Seeking halal food in the US. *Journal of Macromarketing, 36*(4), 400–411. https://doi.org/10.1177/0276146715622243

Keith, N. R., Hemmerlein, K. A., & Clark, D. O. (2015). Weight loss attitudes and social forces in urban poor Black and White women. *American Journal of Health Behavior, 39*(1), 34–42. https://doi.org/10.5993/AJHB.39.1.4

Kornwitz, J. (2017, April 19). *App allows students to find, share food at Northeastern.* http://news.northeastern.edu/2017/04/19/new-app-allows-students-to-find-share-food-at-northeastern/

Larson, N. I., Story, M. T., & Nelson, M. C. (2009). Neighborhood environments: Disparities in access to healthy foods in the U.S. *American Journal of Preventive Medicine, 36*(1), 74–81. https://doi.org/10.1016/j.amepre.2008.09.025

Loftin, J. (2013). *Getting help when needed: Food insecurity among college students and the impact of food pantry availability.* https://stars.library.ucf.edu/honorstheses1990-2015/1426

Loofbourrow, B. M., & Scherr, R. E. (2023). Food insecurity in higher education: A contemporary review of impacts and explorations of solutions. *International Journal of Environmental Research and Public Health, 20*(10), 5884. https://doi.org/10.3390/ijerph20105884

Maguire, J. S. (2016). Introduction: Looking at food practices and taste across the class divide. *Food, Culture & Society, 19*(1), 11–18. https://doi.org/10.1080/15528014.2016.1144995

Martinez, S. M., Frongillo, E. A., Leung, C., & Ritchie, L. (2018a). No food for thought: Food insecurity is related to poor mental health and lower academic performance among students in California's Public University System. *Journal of Health Psychology.* https://doi.org/10.1177/1359105318783028

Martinez, S. M., Webb, K., Frongillo, E. A., & Ritchie, L. D. (2018b). Food insecurity in California's Public University System: What are the risk factors? *Journal of Hunger & Environmental Nutrition, 13*(1), 1–18. https://doi.org/10.1080/19320248.2017.1374901

McArthur, L. H., Ball, L., Danek, A. C., & Holbert, D. (2017). A high prevalence of food insecurity among university students in Appalachia reflects a need for educational interventions and policy advocacy. *Journal of Nutrition Education and Behavior, 50*(6), 564–572. https://doi.org/10.1016/j.jneb.2017.10.011

McArthur, L. H., Fasczewski, K. S., Wartinger, E., & Miller, J. (2018). Freshmen at a university in Appalachia experience a higher rate of campus than family food insecurity. *Journal of Community Health, 43*(5), 969–976. https://doi.org/10.1007/s10900-018-0513-1

Meza, A., Altman, E., Martinez, S., & Leung, C. W. (2019). It's a feeling that one is not worth food: A qualitative study exploring the psychosocial experience and academic consequences of food insecurity among college students. *Journal of the Academy of Nutrition and Dietetics, 119*(10), 1713–1721. https://doi.org/10.1016/j.jand.2018.09.006

Mirabitur, E., Peterson, K. E., Rathz, C., Matlen, S., & Kasper, N. (2016). Predictors of college-student food security and fruit and vegetable intake differ by housing type. *Journal of American College Health, 64*(7), 555–564. https://doi.org/10.1080/07448481.2016.1192543

Moore, L. V., & Diez Roux, A. V. (2006). Associations of neighborhood characteristics with the location and type of food stores. *American Journal of Public Health, 96*(2), 325–331. https://doi.org/10.2105/AJPH.2004.058040

Morland, K., Wing, S., Roux, A. D., & Poole, C. (2002). Neighborhood characteristics associated with the location of food stores and food service places. *American Journal of Preventive Medicine, 22*(1), 23–29. https://doi.org/10.1016/S0749-3797(01)00403-2

MSU student food bank. (2019). *About the food bank.* http://foodbank.msu.edu/about/index.html

Neff, R. A., Palmer, A. M., McKenzie, S. E., & Lawrence, R. S. (2009). Food systems and public health disparities. *Journal of Hunger & Environmental Nutrition, 4*(3–4), 282–314. https://doi.org/10.1080/19320240903337041

Paola, J., & DeBate, R. (2018). Employing evaluation research to inform campus food pantry policy. *Health Behavior and Policy Review, 5*(4), 83–94. https://doi.org/10.14485/HBPR.5.4.9

Patton-López, M. M., López-Cevallos, D. F., Cancel-Tirado, D. I., & Vazquez, L. (2014). Prevalence and correlates of food insecurity among students attending a midsize rural university in Oregon. *Journal of Nutrition Education and Behavior, 46*(3), 209–214. https://doi.org/10.1016/j.jneb.2013.10.007

Payne-Sturges, D. C., Tjaden, A., Caldeira, K. M., Vincent, K. B., & Arria, A. M. (2018). Student hunger on campus: Food insecurity among college students and implications for academic institutions. *American Journal of Health Promotion, 32*(2), 349–354. https://doi.org/10.1177/0890117117719620

Price, C. E., Sampson, N. R., Reppond, H. A., Thomas-Brown, K., & Camp, J. K. (2019). Creating a community of practice among college campus food pantry directors in Michigan. *Journal of Community Practice, 27*(1), 96–109. https://doi.org/10.1080/10705422.2019.1580655

Raja, S., Ma, C., & Yadav, P. (2008). Beyond food deserts: Measuring and mapping racial disparities in neighborhood food environments. *Journal of Planning Education and Research, 27*(4), 469–482. https://doi.org/10.1177/0739456X08317461

Reppond, H. A., Thomas-Brown, K., Sampson, N. R., & Price, C. E. (2018). Addressing food insecurity in college: Mapping a shared conceptual framework for campus pantries in Michigan. *Analyses of Social Issues and Public Policy, 18*(1), 378–399. https://doi.org/10.1111/asap.12161

Rickrode-Fernandez, Z., Kao, J., Lesser, M. N., & Guess, K. (2021). Implementation of a healthy food and beverage policy at a public university. *Journal of Nutrition Education and Behavior, 53*(10), 891–899. https://doi.org/10.1016/j.jneb.2021.06.009

Twill, S. E., Bergdahl, J., & Fensler, R. (2016). Partnering to build a pantry: A university campus responds to student food insecurity. *Journal of Poverty, 20*(3), 340–358. https://doi.org/10.1080/10875549.2015.1094775

Vilaro, M. J., Colby, S. E., Riggsbee, K., Zhou, W., Byrd-Bredbenner, C., Olfert, M. D., Barnett, T. E., Horacek, T., Sowers, M., & Mathews, A. E. (2018). Food choice priorities change over time and predict dietary intake at the end of the first year of college among students in the US. *Nutrients, 10*(9). https://doi.org/10.3390/nu10091296

Waters-Bailey, S., McGraw, M. S., & Barr, J. (2019). Serving the whole student: Addressing nonacademic barriers facing rural community college students. *New Directions for Community Colleges, 2019*(187), 83–93. https://doi.org/10.1002/cc.20372

van Woerden, I., Hruschka, D., & Bruening, M. (2018). Food insecurity negatively impacts academic performance. *Journal of Public Affairs.* https://doi.org/10.1002/pa.1864

Wolfson, J. A., Insolera, N., Laska, M. N., & Leung, C. W. (2024). High prevalence of food insecurity and related disparities among US College and university students from 2015–2019. *Journal of Nutrition Education and Behavior, 56*(1), 27–34. https://doi.org/10.1016/j.jneb.2023.10.013

Wood, J. L., & Harris, F., III. (2018). Experiences with 'acute' food insecurity among college students. *Educational Researcher, 47*(2), 142–145. https://doi.org/10.3102/0013189X17752928

Wooten, R., Spence, M., Colby, S., & Steeves, E. A. (2019). Assessing food insecurity prevalence and associated factors among college students enrolled in a university in the Southeast USA. *Public Health Nutrition, 22*(3), 383–390. https://doi.org/10.1017/S1368980018003531

CHAPTER 7

Campus Food Security: Perspectives from the Pandemic

Abstract Post-pandemic food insecurity has emerged as a significant concern, reshaping perspectives on food access and systemic inequalities. The COVID-19 pandemic exacerbated existing vulnerabilities, highlighting the fragility of food systems and the disproportionate impact on marginalized communities. Many individuals faced job loss, reduced income, and increased food prices, leading to heightened food insecurity. As societies navigate recovery, there is a renewed emphasis on building resilient food systems that prioritize equitable access to nutritious food. Discussions now center on integrating community-based solutions, while addressing the root causes of food insecurity. The pandemic has underscored the need for comprehensive policies that support food security as a fundamental human right, fostering a more equitable future for all.

Keywords Student success • International perspectives • Europe • France • Spain • Africa • South Africa

The University of Minnesota surveyed parents during the pandemic reporting that the pandemic resulted in job loss for 40% of the respondents, of whom 20% stated has negatively impacted their families' finances significantly influencing their ability to assist their student with college-related expenses. Cai et al. (2022) and Stebleton et al. (2020) revealed

S. Sharififard, *Community, Food Insecurity, and a Global Perspective on Campus Food Pantries*,
https://doi.org/10.1007/978-3-031-78223-7_7

that many college students experienced a shift in food security after the onset of Covid-19, which may be related to the closure of college campuses, changes in housing situation, or changes in employment. Because college students may be missing out on economic and nutrition assistance during the pandemic (Mialki et al., 2021). Administrators should also remain up to date with current college food security legislation and make it a priority to assist students as relief becomes available. In 2023, Brown University launched kosher, halal, and allergen-aware spaces in the largest dining hall in Brown's campus as kitchens to provide fresh and accessible meals for students who observe Jewish or Islamic dietary laws, or who have food allergies or other specific dietary requirements (Spear, 2023). There is significance in the cultural and spiritual awareness on campus that students can immerse themselves in an experience that helps their social relations and remove part of cultural barriers where students have what they need to succeed without feeling that they must have a burden of addressing their basic food needs before achieving an inclusive measure of social engagement and belonging to succeed as much as possible.

Several high schools and colleges began piloting programs to include Halal and Kosher lunches in their school districts such as the Los Angeles Unified School District, New York City Department of Education, and Paterson Public Schools. Unless there are partnerships or privately funded ways to fund the needs of students as part of student success initiatives, centering the student voice, administrators must also recognize the aspect of policy where public funds are to be used for private consumption. It is also important to assess the autonomy of law as it relates to activities of public and private charity and how inclusivity addresses the projects that need to be funded versus those that are established at the institution.

In a longitudinal study of food insecurity (Hammelman et al., 2022), 96.4% of respondents were found to be food insecure. 33.57% reported anxiety and uncertainty regarding the household food supply, while 57.8% of people felt that they had to consume insufficient quality food. In 2021, in the Indian state of Kerala, the local government enlisted women in an entrepreneurial network to manage 1,000 temporary hotels to provide over 70,000 subsidized meals every day because of the global pandemic (Nair & Muley, 2024). The women persistently endured severe levels of food insecurity as they helped their local community. Achieving gender equality and women's empowerment is integral to each of the 17 SDGs with SDG 5, Gender Equality, supporting the economic status of women. The social and financial perspective in terms of its mobilization of women has often prevented women from completing a formal education, going to school, and reaching their potential and skillset. Instead, their educational pursuits have been integrated with an emphasis on community and the social safety net to decide the best approach in acquiring resources. Interestingly, students and families integrated food safety nets as a community endeavor (Kerala, 2023) that also had a greater mental health impact. Whereas individuals of different faiths would tend to avoid each other's food—someone on the network would not have questioned how the food was made, what religion or race of the food being made, and no identity markers attached. The woman cooking the meals had no idea who the beneficiaries were. People accept it gratefully as food, just food.

In the US, similar pandemic-related programs shifted the cultural narrative facilitated in collecting food, cooking food, and creating a sustainable food system. When the food was known to be organic, the stigma was reduced overall. Although these findings are not consistent with other places that have food pantries, having organic produce available could help with decreasing feelings of isolation and shame related to food assistance. These implications contribute to the moderating role of food and consumer behavior that contributes to one's overall health in a positive way instead of perpetuating food insecurity and environmental concerns from a public health perspective and socioeconomic outcomes. Recent research has suggested that the consumption of organic foods contributes to prosocial behaviors, physical fitness, and environmental benefits (Tomar et al., 2023). The consumption of organic food, because better intake leads to a better thought process and eventually the results get reflected in the form of a better human being. To conclude, better thoughts, emotions, mind, and health of a human individual and society are strongly rooted in the adoption of organic foods. One can be conscious of their food consumption and make choices that benefit themselves and their communities. However, if they have negative effects on their academic performance, professional and personal responsibilities, the positive attributes of the organic food consumption can be negated and less frequently associated with positive behavior and dietary patterns overall.

Barcelona, Spain, has 39 public food markets spread across the city (Barcelona, 2021) serving every neighborhood and population respectively. The government funds the markets while supporting digital infrastructure and public health outcomes related to equitable food systems that represent a student and community population, which includes providing Kosher and Halal foods. The quality of foods at the market brings positive impacts to the convergence of public health concerns and risks associated with academic and socioeconomic disparities. Currently, they offer culturally competent services and refer client to culturally relevant programs and services in Barcelona, Madrid, and Mallorca. Staff are culturally diverse with intentional hiring practices in place and a focus on accessibility and inclusivity. Uniquely, translation services are also available for patrons (González-Torre & Coque, 2016).

In Brazil, 45% of emergency food supply initiatives were based on local supply chains, and over 85% were at least partly agroecological. To encourage and prepare stakeholders, these initiatives were scaled across to benefit the emergency food response system irrespective of need. The existing literature offers limited insight into how food pantries help students in need of food assistance. Much of the literature suggests that students experiencing food insecurity have been facing other academic and personal challenges while navigating their individual college experiences. The literature further suggests that campus food pantries are just one solution to different ways to approaching the issues surrounding food insecurity on campus.

References

Ajuntament de Barcelona. (2021). *The City Council presents the next investments in the municipal markets.* https://ajuntament.barcelona.cat/bombers/en/noticia/the-city-council-presents-the-next-investments-in-the-municipal-markets_1055163

Cai, Q., LeBouef, S., Savage, M., & Dworkin, J. (2022). What happened when COVID-19 shut down in-person higher education? Parents speak out. *About Campus: Enriching the Student Learning Experience, 26*(6), 26–34. https://doi.org/10.1177/10864822221082695

González-Torre, P. L., & Coque, J. (2016). How is a food bank managed? Different profiles in Spain. *Agriculture and Human Values, 33*, 89–100. https://doi.org/10.1007/s10460-015-9595-x

Government of Kerala. (2023). *Rs 33.6 Crore Subsidy for Kudumbashree Janakeeya Hotels; 5043 entrepreneurs of 1198Janakeeya Hotels benefitted.* Local Self Government Department. India. http://lsgkerala.gov.in/index.php/en/kudumbashree/news/rs-336-crore-subsidy-kudumbashree-janakeeya-hotels-5043-entrepreneurs

Hammelman, C., Shoffner, E., Cruzat, M., & Lee, S. (2022). Assembling agroecological socio-natures: A political ecology analysis of urban and peri-urban agriculture in Rosario, Argentina. *Agriculture and Human Values, 39*(1), 371–383. https://doi.org/10.1007/s10460-021-10253-7

Mialki, K., House, L. A., Mattews, A. E., & Shelnutt, K. P. (2021). COVID-19 and college students: Food security status before and after the onset of a pandemic. *Nutrients, 13*(2), 628. https://doi.org/10.3390/nu13020628

Nair, D., & Muley, A. (2024). Food security and dietary diversity patterns amongst female handloom workers: A cross-sectional study from Kerala, India. *Discover Food, 4*(23). https://doi.org/10.1007/s44187-024-00091-x

Spear, M. (2023). *New kosher, halal and allergen-aware kitchens serve up inclusivity, nourishment at Brown*. News at Brown. https://www.brown.edu/news/2023-09-21/kitchens

Stebleton, M. J., Lee, C. K., & Diamond, K. K. (2020). Understanding the food insecurity experiences of college students: A qualitative inquiry. *The Review of Higher Education, 43*(3), 727–752. https://doi.org/10.1353/rhe.2020.0005

Tomar, S., Sharma, N., & Kumar, R. (2023). Effect of organic food production and consumption on the affective and cognitive well-being of farmers: Analysis using prism of NVivo, etic and emic approach. *Environment, Development and Sustainability*, 1–22. https://doi.org/10.1007/s10668-023-03195-z

United Nations. (2015). *Sustainable development goals*. Department of Economic and Social Affairs, Sustainable Development. https://sdgs.un.org/goals

CHAPTER 8

Ethnic Foods Among Neighborhood Characteristics

Abstract Ethnic foods play a significant role in shaping neighborhood characteristics, contributing to cultural identity and community cohesion. In diverse urban areas, the presence of ethnic restaurants and grocery stores fosters social interaction and creates vibrant culinary landscapes. These establishments not only serve as sources of authentic cuisine but also attract residents and visitors, enhancing local economies. Additionally, ethnic foods can influence neighborhood demographics, often reflecting the cultural heritage of the community. The availability of diverse food options promotes inclusivity and enriches the cultural fabric of neighborhoods. However, it can also lead to challenges such as gentrification, where rising property values threaten the preservation of cultural spaces. Understanding these dynamics is essential for promoting sustainable, inclusive development in multicultural communities.

Keywords Grocer • Food retailer • Food manufacturing • Supply chain • Migration • Fast-food • Convenience stores

Disadvantaged neighborhoods with smaller grocers need more access to nutritious foods (Ohri-Vachaspati et al., 2019). Residents have a longer commute to larger supermarkets with fresh produce than in predominantly White neighborhoods. The need for more availability of retail stores

S. Sharififard, *Community, Food Insecurity, and a Global Perspective on Campus Food Pantries*,
https://doi.org/10.1007/978-3-031-78223-7_8

affects the lack of grocers and businesses offering quality produce in these neighborhoods. Areas with higher proportions of African-American/Black residents have more fast-food restaurants and fewer supermarkets than neighborhoods with higher proportions of White residents (Sansom & Hannibal, 2021). Non-fast-food restaurants are more prevalent in racially mixed and White neighborhoods than in African-American/Black or Hispanic neighborhoods, influencing the diet and food quality of individuals residing there. Other researchers have found no difference in diet quality because of living closer to a supermarket but have considered food environment measures for neighborhood characteristics. They will identify resources that colleges and universities would benefit from using to create an inclusive and diverse food pantry on campus and enhance students' safe sense of identity, profoundly influencing how students experience higher education. Additionally, how administrators and student success advisors respond to persistent or emerging student concerns is critical to identifying educational policies that support student success outcomes, from when they first enter school to graduation.

Food policy and current food systems are all central to understanding food insecurity, as well as the cultures and populations that are affected within a community. The global food system is interconnected within a global system embedded across non-governmental organizations as well as institutes of higher education. Practicing respect for the food preferences of individuals and dietary needs is essential to understanding how social and political systems affect the community. This study is conducted from a U.S. perspective while recognizing the global food system and is guided by the spatial complexity of organizations. Spatial complexity refers to how an organization is shaped by many dynamics within the food system, changing the local context as impacted through space and time. The Stigma and Food Inequity Framework suggests that there are structural and individual levels of stigma that are mediated by different factors, including access to resources, food environment, and psychosocial and behavioral processes. According to Martin et al. (2021), when food pantries in New England selected healthier foods after nutrition information was available on their food bank's ordering platform, the overall nutrition environment of food pantries improved. The selection of more nutritious foods remained. These findings suggest that pantry staff would be willing to make better decisions for their patrons based on the information they have, irrespective of cost and dietary restrictions.

Intercultural and Interfaith Food

Foods identified as Halal, Jain, and Kosher are not commonly found across campus food pantries or in the surrounding municipality. These foods are accepted, encouraged, and permitted under dietary law. Since these foods and products must meet specific guidelines, a structured system must ensure appropriate certification and regulation. Certification acknowledges that an organization, food service provider, or distributor has followed all necessary procedures and preparations to ensure that all their products or items are lawful and permitted under its specific law. Certification is also essential because it assures consumers that there is no question about the purity of the product or food. Specifically, students who abide by these dietary guidelines might only purchase foods and items they know are guaranteed to be permitted under the law as per their faith or preference. Jurisprudence differentiates distinct categories of food, where some foods, specifically meats, are categorized as permitted or forbidden. This affects the production process of foods, the source of consumption, and how the foods are handled, packaged, and transported across the global supply chain. It also impacts how students, consumers, and individuals shop for food and choose where to live, work, and study. While legal and religious scholars often debate the interpretation of the permissibility of certain food products and their production, food consumption has commonly been defined in public health settings. Namely, the shared features include:

- high hygiene standards,
- the quality and cleanliness of ingredients,
- processing methods,
- the cleansing and ethics of human and animal rights, and
- concerns for safety and environmental wellness.

These definitions and classifications have recently expanded and impacted the global food system to include various products across different markets. The sustainability and quality of such foods have also become increasingly difficult to determine as their characteristics have evolved from communities where food buying patterns have influenced modernization and assimilation. For example, international students are adjusting to campus and seeking foods that meet their dietary needs. When the students are still looking for food that meets the accepted certification of the food product desired, it can become difficult to integrate into their new communities socially. This impacts their studies and the quality of completing and managing their course schedule with their professional and personal commitments.

Consequently, it is critical for students who are beginning to understand their community, culture, and preferences for preparing, acquiring, and storing food. Understanding the available meal plans and dining options on and off campus can help them avoid encountering significant challenges in finding food products that meet their needs in the long term. This impacts how they sustain their diets and change their food consumption patterns over time. The food system changes in response to consumer demands and is affected by individuals who follow specific food customs over time. One should consider terms and conditions when shopping for food without compromising the other, especially regarding dietary customs and food preparation (Martin, 2021).

The spatial complexity of the food environment can inform how the community food environment contributes to better and preferable consumer interests over time. Dietary preferences and food laws might be more conducive to secular food rules or preferences when operating food banks that serve a diverse population. It is an additional consideration to infer what happens when religious/secular relationships begin to cause shifts in public institutions.

Additionally, cross-contamination risks associated with acquiring food products are unknown, as only sometimes visible. There is no guarantee of knowing whether the food has been certified by an authorized body or has the appropriate handling. However, specific mitigation tools or cleansing processes reduce the risks over time. For example, a cleansing ritual can be performed when handling food pantries and dining halls, where there is the possibility of visible cross-contamination of food products.

The maximum level of compliance with the guidelines would be reached when all the utensils, handling equipment, and tools are correctly sorted, cleansed, and used separately in both locations where they are stored. It is critical to carry this practice over during the distribution. If labeling is used, there will be issues with using the materials in the same area. Further, proper disposal of these materials would be required, and would be the use of safe and sustainable ways to be evaluated for future use among consumers. Those who manage the food pantries and dining areas would be responsible for ensuring that their students and patrons understand the food labels being used and recognize the food's processes, customs, and applications, such as Kosher and Halal. These practices would be consistent to ensure that students' preferences are respected on their campus.

Further, low-income, minority and immigrant communities disparately affected by food production and restrictions particular to their diets and religious accordance (D'Haene et al., 2019) may also encounter challenges with commuting to an accessible grocer. Such communities' zip codes and proximity to various foods are critical to the types of barriers that may exist in those sectors and their relationships to diet quality and neighborhood safety. It is the time to shape critical dialog on campuses and in the community. It is beneficial to understand how all members of the community access and consume food. Determining how food is acquired, prepared, stored, and handled is also essential. Food banks could help develop culturally relevant partnerships with their communities while influencing communities where individuals requiring such services are valued and emotionally, socially, and culturally connected.

Secondly, volunteers and staff can promote leading opportunities that support a culture of inclusion and well-being as part of their services. These events further open the opportunity to address other community concerns and communicate ideas for promoting wellness and well-being. Food distributions should not stigmatize any community member, volunteer, or staff. If there is also a growing interest in selecting categories from additional choices of foods, it would be beneficial to identify the variety of foods that would be necessary and requested by patrons. If specific food retailers or other grocers are unavailable, individuals might challenge how their community meets their personal and social needs. Geographic, racial, and class disparities and health discrepancies can persist if the interplay between food preferences, social networks, and the shift of neighborhood characteristics over time do not align with specific or recommended dietary preferences, food choices, and the availability of markets and food products.

On average, the food pantries in this study maintained daily operations in an academic year for approximately 18 hours a week, including once a week (n = 3, 5.56%), 2–3 days per week (n = 17, 31.48%), 4–6 days per week (n = 8, 14.81%), weekdays (n = 12, 22.22%), and 7 days per week (n = 3, 5.56%). Further, participants reported having opened their campus food pantry with no continually operating hours, while others reported that their hours of operation are once a week (n = 4, 7.41%), once every other week (n = 3, 5.56%), or once a month (n = 3, 5.56%). One other participant indicated that the hours of operation are based on the number of federal work-study hours available for student service. As the findings originally revealed, about 11% of pantry employees are student workers, mostly funded by the Federal Work-Study Program.

The food pantries were reported to be located throughout their campus with a large number in Student Services/campus centers (*n* = 18, 33.33%) as well as Student/University unions (*n* = 12, 22.22%). Other pantries were reported to be in campus residential housing (*n* = 4, 7.41%) and academic buildings that host academic departments and institutes, student organizations, and classrooms (*n* = 10, 18.52%). Another pantry is in a financial aid office, and one other university food pantry was reported to be in the university dining hall. These locations offer students the opportunity to address their basic needs in one geographic area where there are other colleges/university facilities and activities.

Pantries were further reported to be in academic buildings where there are programs, academic centers, and classrooms. Periodic campus pop-up food distributions, mobile pantries, textbook rental offices, campus police buildings, campus parking lots, and the offices of Deans of Students were common. Food pantries that are in academic buildings have an additional advantage of fulfilling their independent goals by having the chance to work closely with students and faculty in researching food insecurity issues. In addition to food pantries, some participants (n = 2) also reported offering additional student services, such as professional closets, where students could borrow professional apparel for their job interviews.

Communication issues, cultural barriers, and student services are important to understanding the specific needs students desire from campus, as well as the opportunities and activities they perceive as relevant to campus and student services. As one participant shared, cultural misunderstandings and individual interpretations can negatively affect campus food pantries in the following ways:

> We are new, but a common problem among food pantries that we've seen is the cultural differences of international students not understanding what a food pantry is for and taking free food because it's free rather than because they really need it. The explanation of 'what is a food pantry?' sometimes doesn't get picked up because oflanguage barriers. We're changing that this year with a session at international orientation about why the pantry exists, who uses a food pantry in the United States, and the cultural context behind it. The goal is to make sure that graduate international students are eating nutritious food, because we also know that they can be making less money

> than others, have a hard time finding even low paying on campus jobs/internships, and live together off campus for cheaper rent. We have also had stories of international graduate students being on campus and not getting paid and not telling their supervisors because they did not want to be sent away, so they were starving. There are students in a lot of need here, so the communication across cultures aspect is something we're looking into trying to find the best way to communicate first to get the point across. (Participant, 17)

Food insecurity issues are only part of a more significant lack of basic needs affecting students on campus. When asked if the pantry only offers food items, 88.3% of respondents reported that other items are also available in their pantries, including toiletries, personal care, hygiene products, school supplies, cooking and kitchenware, and cleaning products ($n = 48$). Ninety percent of respondents indicated that their food pantries are not a temporary solution, which implies that addressing food insecurity issues in academic environments is a deep and complex issue.

Also, when asked if patrons are primarily students who have used their meal points, 82% reported that is not the case. When asked if there is consistency in the pantry among the number of food items, respondents reported:

Occasionally, 10%
Fairly many times, 15%,
Very often, 15%, or
Always, 10%

Participants further reported meeting food regulations and environmental safety concerns in operating their food pantries. For example, some of the participants (n = 9) reported strictly adhering to accepting donations of unopened goods; and six participants reported that the temperature of the pantry must meet specific guidelines for quality and safety.

In terms of food quality, a few of the participants (n = 4) indicated that donated foods must be unopened or unwrapped, unexpired, and cannot be homemade. Other pantry respondents (n = 7) reported routinely checking expiration dates on items like bread and cheese, while others reported either tossing out or rejecting expired foods that are donated or ready in stock (n = 9).

Two respondents reported that their pantry only offers canned goods, while four other pantries specified that they only accept non-perishable food items.

About 40% of participants indicated they were either operating as an agency of their local food bank or partnering with their local food bank(s). As such, they generally follow either local food bank guidelines or the Feeding America organization, a nonprofit working on hunger and poverty issues that services food banks in the United States. Five percent of the food pantries are functioning under state and local regulations. Twenty-five percent of the food pantries are functioning under their campus' risk management offices. Thirteen percent of the food pantries function under USDA guidelines.

The number of students visiting their campus food pantries has increased overall. Eighty-two percent of participants reported that in an academic year, over 100 students are campus food pantry patrons. This number excludes the total number of individuals who use the pantry, including faculty and staff, as well as others who are permitted to use the pantry's services.

Eighty-five percent of participants reported that the number of student patrons has gone up much higher than in previous years. The need becomes more critical as food pantry administrators begin developing other efforts to sustain the pantries' operations and services. As expressed by a participant:

> Due to the high number of clients in need of food, we are often running out within the first 30 min-1 hr. Because of this high demand of food, a higher volume of food pantries is needed to fully provide food access to all students in need. (Participant, 41)

Interestingly, the study's survey had been distributed to all participants in the fall semester. The effects of food insecurity on campus are immediately recognized across the represented pantries and are increasing from year to year. However, students might have to compromise their food access if the food type they would like is not available at the pantry. About 15% of participants reported that their food pantries have limited operating hours of service. If a student can only visit the pantry at certain times in the day, they will likely not meet their effort in accessing basic food if their campuses' food pantry has limited hours of operation.

In a series of open-ended responses, nine participants reported that their pantry would either assign Case Managers to students who would frequently visit the pantry; or assist those students with filling out an application. Other participants (n = 5) reported having a policy in place for arranging an appointment with a Case Manager. Sixty-two percent of participants indicated a cap on the number of *items a student can take* per visit to the pantry. Thirteen participants reported that their items per visit are limited to once a week.

Five participants reported that a student could visit the pantry twice a month, or twice a week (n = 2), with up to 60 visits per student. Fifty-six percent of participants stated that there is a cap on the number of *visits* per student. Students could take two full grocery bags per visit (n = 2), 10 pounds of food per week, or 9–12 items with only one item per section.

Participants did not report any specific policies affecting how they determine the hours of operation or the number of items per capita overall. One participant confirmed that the school's food pantry does not have any policies, because they are aware that several students at their campus are feeding households with children and dependents. While some pantries (n=4) use a point system to track items per patron, others do not enforce a limit. Because some of the responses to the question related to the cap on the number of *items* per visit included self-reported responses, individual replies varied. Therefore, the researcher could not use those responses for purposes of statistical significance.

However, offering foods of a lower nutritional value and foods with no dietary or cultural contents was not perceived as a direct barrier to overall food access. Instead, the types of food items currently available inside the pantries are necessary for maintaining operations; and follow specific guidelines and regulations. Food donations can often limit the types of food items that might be requested or preferred by a staff member and student patrons. The limitations of some internal policies on operating a campus food pantry might also affect the types of foods that a pantry can offer in an academic year. A campus food pantry staff member might like to offer a variety of select food options and basic needs items in their food pantry. However, the unpredictability of donations poses additional barriers to food assistance.

In specific, donations provided to the pantries included items purchased by pantry staff members, as well as foods brought in through community partnerships. Eighty-four percent of participants reported that the types of food offered at the pantry vary by the donations received. Other participants reported that their donations are consistent with the availability of food items from their partnering food bank (n = 6). Seventy-nine percent of participants reported carrying over 100 available food items in their pantry, while 10% replied that they have between 76–100 available food items.

In self-reported responses, some food participants (n = 4) reported that their pantries include low sodium, sugar-free, and dairy-free food products to meet nutritional recommendations. Additionally, three participants reported that their pantry offers unlimited fresh produce. About 63.5% of respondents indicated that there are no specific policies that would affect how students experiencing food insecurity could better address their needs. However, a few participants (n = 5) reported to providing students with alternatives to helping solve food insecurity issues, such as establishing a community garden on campus, fundraising for food security and basic needs causes, and promoting food safety and wellness across campus student organizations and administrative departments.

While the local landscape of the food environment can be an independent indicator of food insecurity in the community, it might not be generalizable to the whole population. The researcher believed that embedding the case study analysis with the quantitative aspects of the study provides a better understanding of the barriers to food access as well as categorically different food options. The physical environment and the various food donations and food pantry partnerships on-campus that pantry leaders would encounter with their current food environment would also be discussed on a broader community level.

For the purposes of this study, it is essential to understand the neighborhood context as well as the educational opportunities in the area. There have also not been any case studies examining specific combined neighborhood demographics, spatial complexity, and food accessibility from a higher education perspective. Yet, these factors are important in determining some of the likely neighborhood characteristics affecting food insecure populations on college and university campuses.

In general, food deserts are often found in abandoned communities or vacant homes; and have smaller populations. Residents tend to have lower levels of education, earn lower incomes, and have higher unemployment rates. Food desert tracts comprise different ethnic and racial compositions across the U.S. Based on the research, it is suggested that individuals living in food desert areas have low access to healthful foods and are more likely to experience food and nutrition insecurity.

Such findings are further important to understanding educational settings and the neighborhood food environment, especially for students who intend to enroll in local colleges and universities after completing their precollegiate education. Demographic factors related to housing status, early childhood education, academic achievement, and varying levels of socioeconomic status are also critical to understanding food insecurity across college campuses.

Ethnic markets, organic and whole, and natural foods stores were limited in each county. There were more grocery stores than any other type of food retailer in the counties. No direct relationship could be discerned from each county's nutrition insecurity among food deserts.

Food store accessibility represents the distance to the nearest supermarket, convenience store, or grocer, for a resident living in the center of the county or neighborhood. According to the USDA Food Environment Atlas, there are about five traditional full-service and fast-food restaurants available per one thousand people in the lowest areas experiencing severe poverty and food insecurity. The average household spends 15.9% of its expenditure on food necessities (Volpe et al., 2024).

Areas, where multiple colleges and universities serve large student populations, can strengthen connections between local institutions and create new opportunities past the challenges of acquiring food and emerging food insecurity. By identifying how campuses are identified and represented at the community level, technological advancements can further help public institutions attract more attention toward the cycle of food manufacturing, food products, food retailing, and food insecurity. Based on the case study analysis, convenience stores are more frequently in food deserts and low-income urban areas than the rest of the U.S. While a similar trend was not found globally, there were no significant relationships among grocery stores, specialized food stores, and small and large grocery stores.

References

D'Haene, E., Desiere, S., D'Haese, M., Verbeke, W., & Schoors, K. (2019). Religion, food choices, and demand seasonality: Evidence from the Ethiopian milk. *Foods, 8*(5), 167. https://doi.org/10.3390/foods8050167

Martin, K. S. (2021). *Reinventing food banks and pantries: New tools to end hunger.* Island Press.

Martin, K., Xu, R., & Schwartz, M. B. (2021). Food pantries select healthier foods after nutrition information is available on their food bank's ordering platform. *Public Health Nutrition, 24*(15), 5066–5073.

Ohri-Vachaspati, P., DeWeese, R. S., Acciai, F., DeLia, D., Tulloch, D., Tong, D., Lorts, C., & Yedidia, M. (2019). Healthy food access in low-income high-minority communities: A longitudinal assessment-2009–2017. *International Journal of Environmental Research and Public Health, 16*(13), 2354. https://doi.org/10.3390/ijerph16132354

Sansom, G., & Hannibal, B. (2021). Disparate access to nutritional food; Place, race and equity in the United States. *Nutrition, 7*(29). https://doi.org/10.1186/s40795-021-00434-2

Volpe, R. J., Cai, X., Roldan, P., & Stevens, A. (2024). Food prices in the United States during COVID-19: Generalized facts on price inflation and volatility. *British Food Journal.* https://doi.org/10.1108/BFJ-05-2023-0421

CHAPTER 9

Food Systems and Policy Implications

Abstract Food systems are complex networks that encompass the production, distribution, consumption, and waste of food, directly impacting public health, environmental sustainability, and economic stability. Effective food policy is essential for addressing challenges such as food insecurity, malnutrition, and climate change. Policymakers must consider the interconnectedness of local and global food systems to create equitable and resilient frameworks. This includes promoting sustainable agricultural practices, enhancing access to nutritious food, and supporting small-scale farmers. Additionally, policies should focus on reducing food waste and improving food safety standards. By integrating stakeholder perspectives into policy development, food systems can meet immediate needs and promote long-term sustainability and community well-being. These implications highlight the necessity for collaborative, informed approaches to food governance.

Keywords Sustainability • Inclusive food community • Production • Distribution • Unintended consequences • Property

The food system has direct effects on individual health and food consumption. Food consumption, as determined by food quality, quantity, and safety, affects the food system indirectly. A sustainable food system is one

S. Sharififard, *Community, Food Insecurity, and a Global Perspective on Campus Food Pantries*,
https://doi.org/10.1007/978-3-031-78223-7_9

that provides foods that are accessible, acceptable, adequate, and available (Shannon et al., 2015). Some of the changes that are shaped by the local food context include environmental shifts, such as pollution. These changes can, in turn, create indirect and unintended consequences for the overall food system.

Food prices and food insecurity rely on several assumptions about the relationship between costs and consumption, as well as consumption and food security. Food production and productivity respond to technology and the availability of land for irrigation. Consumer income and food preferences influence the demand for specific foods. According to the National Research Council (2015), as incomes rise, consumers demand more animal protein sources, which generally require higher prices of production. External forces such as weather, transportation, agricultural policies, food processing, food marketing, food packaging, and restrictions of various retail sectors have also impacted consumer food prices in the food supply chain. However, food consumption patterns are not more likely to significantly differ or change among populations.

Small changes in food prices can significantly affect food consumption patterns. Focusing individually on specific consumer behaviors and consumption patterns to the changes in food prices can be problematic. Food production and consumption vary across regions and populations. Differences in price, food preferences, and income are all part of the ways in which consumers change their food consumption and make decisions about their purchases. The per capita consumption of the same or different foods also varies among populations across the U.S. Consumers continually make different dietary choices, whether conscious, about food consumption within a broader context of available, affordable, and acceptable foods (Nesheim et al., 2015; Tom et al., 2016).

Additionally, the core aspects of the U.S. food system, as it relates specifically to fruits and vegetables, include several environmental outcomes and external processes that affect the agencies involved in the supply chain. To follow the USDA dietary and nutritional recommendations, one would have to significantly increase their overall fruit and vegetable consumption (Krebs-Smith & Kantor, 2001). The individual quantity of such consumption would affect the entire food supply chain and potentially cause threats to the food system over time. Further, prolonged dietary intake consistent with similar nutritional recommendations would ultimately affect the relationship between the environment and dietary patterns.

Understanding some of the systemic, legal, physical, and attitudinal barriers or restrictions that student populations face in making food choices is an important aspect of learning about food-related and dietary consumption beyond spaces in higher education. It is also important to consider the best needs for students and patrons, such as how students with specific dietary needs continue to alter or improve some aspects of their regular diets without risking the quality of the foods they are used to consuming. Building a pantry that continues to reflect student needs is important for various reasons, including improving student services, responding to student concerns, and creating a place for students to share the benefits of community wellness. Students can also learn more about the global food system and its impact on individual and local communities.

At an organizational level, educational leaders could better understand the need for systems thinking (Senge & Sterman, 1992) in their services related to food to benefit their respective institutions, and support campus food pantries to include new or additional food items in the pantry. The importance of higher education institutions understanding the global food system and the diverse student populations in the U.S. that impact student enrollment can also impact departments on campus, the types of donations and foods available on campus, and the types of academic and support services programs across an institution. Systems thinking applications could aid in developing more equitable and inclusive food systems at institutes of higher education.

Moreover, the aspirations for attending a school that meets a student's needs can impact how students will perform in the classroom and at home. Educational leaders who oversee their campus food pantries would likely benefit from being informed about food items that are best suited for their students. They can also consider ways to identify different types of foods that are best to serve the needs of diverse student cultures and dietary needs. In turn, students could have the ability and resources to store and consume food more safely and comfortably without feeling shame or discomfort. This impacts the whole student development and how they measure their individual growth and practical skills training.

First, developing a college or university task force can help with piloting different programs based on identifying individual student-centered food needs and other essential needs, such as school supplies, personal hygiene products, and cooking utensils and equipment. If a task force has already served campus food pantry and food insecurity projects, campuses can consider the benefits of identifying existing state and local policies that affect how campus operations are more broadly defined at an institutional level, and the direct impacts of their communities' greatest needs. For example, the critical role that colleges and universities have in admitting students who have the goals and willingness to complete their education to the highest levels will leading to higher graduation rates, and employment opportunities that can help prevent the generational cycle of poverty. Inclusive economic growth begins with colleges with the support they provide to help them benefit from their program rather than perpetuate economic inequality and hardship.

The task force can consider using internal communications and promote similar initiatives of the pantry to several different institutional programs or public or nonprofit organizations that are closely aligned with the same mission and values to help advance student success and wellbeing. Students can be provided with options to choose a path that is consistent with their program learning outcomes and get involved with student groups and professional networks to expand their pool of opportunity as a student. Students could also be encouraged to work with student employment and human resources departments to boost morale, emphasize collective campus and student needs, and promote public and community support in the classrooms.

Human resources and talent management offices can help their employees interpret shared meanings of student needs and access to different student resources across campus departments.

Next, the task force can seek out opportunities to gather confidential, anonymous basic needs information from student perspectives on a semester or quarterly basis. To understand some of the changing dynamics of academic wellbeing and student success, it might be most beneficial to distribute the surveys twice a year, occurring at the beginning and end of the academic school year or as appropriate for the institution's needs. Some of the results could be used to infer meanings about strategies to implement in future terms. The results could also be used to identify persistent forms of food insecurity experiences, as well as ongoing basic needs. Survey distributions could potentially occur during Orientation programs for first-year undergraduate/graduate students, and subsequent events where other signature campus events take place. Using basic personalization of survey invitations might be used to encourage students to feel welcome to provide feedback while remaining assured that their information and responses are confidential and anonymous. Digital distribution is also an important factor to consider when understanding the benefits and advantages available to students who would like to participate when managing their academic or professional schedules.

The task force can then consider opportunities for student check-ins throughout the academic term. Communicating and creating spaces where students are comfortable in addressing their concerns and using check-in facilitations on-campus or in virtual settings can be an important aspect of the learning environment and help with creating spaces for meeting with students. If there are students who live off-campus, the task force can work to offer students virtual check-in opportunities or other appointments, as necessary. Over time, it might be valuable for staff and faculty members at different departments to conduct student wellness checks on a rotational basis. Such communications can allow staff and faculty to better understand some direct insight into student concerns and the overall campus environment. Source culturally appropriate food as much as possible and be more conscious of the respective needs of what is acceptable and appropriate in developing an inclusive food-based community around student success. Destigmatize food that looks, tastes, and is prepared differently.

With respect to hosting food distributions on campus, the task force can consider the positive influences of including more nutritious and inclusive foods as identifiable among the student population. While the primary purpose of the task force might be to engage students with using campus services and identifying urgent student needs, it would be valuable for students to have the assurance and encouragement that their requests and essential needs have been addressed with supportive staff and faculty. Also, educating students and food pantry staff members about nutrition security and nutritious food content is important to advancing the understanding and utility of having a sustainable campus food environment.

Additionally, hosting workshops and webinars to introduce events about getting involved with the campus food pantry and encouraging volunteerism are helpful ways to get students to engage with the committee's work. Students could be encouraged to participate in sessions about food preparation and management, access to nutrition, and the retail food environment virtually. Examples of activities can include community health speaking series, academic success newsletters with tips for academic success, and wellness and basic needs webinars.

Further, course instructors can include basic needs information in their course syllabi, consistent with school policies. Action-based research projects can also attract undergraduate and graduate students involved with the school's community—bringing more cross-departmental work across policy, public health, organizational behavior, education, and psychology researchers. As an example, students enrolled in legal studies programs can work with students in public health and nutrition programs to better understand the extremity of the issues surrounding food insecurity and identify legislative implications at the local and federal level.

Having the opportunity to integrate coursework with field experiences and identifying outreach and administrative efforts can increase participation among students from various institutional and academic programs. Additionally, students enrolled in fully online programs or who live off-campus or who do not have reliable transportation would have more opportunities to participate in such efforts and might have more ways to initiate dialog with the institution and their peers.

The task force members and food pantry staff could then consider technological alternatives to helping students with meal preparation services to access the foods they desire. While food delivery services and digital meal programs do not necessarily affect the availability of transportation or vehicle access for students, such programs can help food pantry staff to purchase food items for their pantries that are necessary and required for student wellness. The staff could also provide periodic customized programs for students to select options that they would like to have on an individual or item-by-item basis. Applying technologies to food access problems can also help with reducing food waste; and acquiring individual and necessary foods. In turn, students might also have more flexibility to participate in professional development events and extracurricular activities; and not experience increased stress with planning their meals.

Students could further have the option to use vouchers for meals not offered at the pantry or within the campus food environment. Students can then obtain what would normally prohibit one from participating in a campus plan or food pantry. However, there might also be opportunities for students to work with other students from different programs and concentrations to promote departmental activities where they collaborate to find appropriate and timely solutions for their peers and colleagues.

Further, food pantry staff members can help develop culturally relevant partnerships with the community while establishing other relations within different departments on campus to ensure that all students are valued and respected emotionally, socially, and intellectually throughout their academic studies. School functions such as orientation, community building sessions, and student-centered workshops are opportunities for students to create a culture of inclusion and wellbeing. These events can further open the opportunity to address student concerns and communicate ideas for student success initiatives such as the growth of campus food pantries.

To comply with food and dietary rules/customs, the pantry staff could work to ensure appropriate restrictions of cross-exposure of certain food items. In that same respect, food distributions should not stigmatize any student or pantry representative. Instead, it might be advantageous to conduct multiple food distribution events throughout the year. Additionally, pantry staff members could work with community food vendors and meal delivery services to apply discounted rates and student offers to integrate within the campus food pantry. Experimenting with different policy-based and marketing proposals could allow the pantry staff members to determine what would be best for the students and the campus community.

For student workers who serve in the campus food pantry, there is a more focused awareness of how students can respond to the different food choices on campus. The student employees can advise staff members on ways to best approach student needs and grow the camaraderie among their peers in a collegial environment. If there is also a growing interest in selecting categories from additional choices of foods, it would be beneficial to identify the variety of foods that would be necessary and requested by students.

College and university admissions and enrollment officers also could consider how they respond to persistent or emerging student concerns as they admit prospective students. Admissions representatives can assess whether an incoming class of students with environmental and ecological interests surrounding a campus would consider attending their school on those types of services and activities. If specific food retailers or other grocers are neither physically on campus nor in the community, several students might begin to develop concerns about how the college or university would best serve their personal needs with their academic goals.

Members of admissions offices can also assess whether international students would choose to go to a school that is more affordable for their financial means over a specific type of international and related multiethnic and multicultural community that would allow them to thrive more socially and academically. College applicants also have many choices to consider when selecting the college or university that will give them the best educational opportunity and experience.

References

Krebs-Smith, S. M., & Kantor, L. S. (2001). Choose a variety of fruits and vegetables daily: Understanding the complexities. *The Journal of Nutrition, 131*(2), 487S–501S. https://doi.org/10.1093/jn/131.2.487S

National Research Council. (2015). *A framework for assessing effects of the food system.* National Academies Press.

Nesheim, M. C., Oria, M., & Yih, P. T. (2015). *A framework for assessing effects of the food system.* National Academies Press.

Senge, P. M., & Sterman, J. D. (1992). Systems thinking and organizational learning: Acting locally and thinking globally in the organization of the future. *European Journal of Operational Research, 59*(1), 137–150. https://doi.org/10.1016/0377-2217(92)90011-W

Shannon, K. L., Kim, B. F., McKenzie, S. E., & Lawrence, R. S. (2015). Food system policy, public health, and human rights in the United States. *Annual Review of Public Health, 36*, 151–173. https://doi.org/10.1146/annurev-publhealth-031914-122621

Tom, M. S., Fischbeck, P. S., & Hendrickson, C. T. (2016). Energy use, blue water footprint, and greenhouse gas emissions for current food consumption patterns and dietary recommendations in the US. *Environment Systems and Decisions, 36*(1), 92–103. https://doi.org/10.1007/s10669-015-9577-y

CHAPTER 10

Conclusion

Abstract Food systems in higher education are crucial for promoting student health, sustainability, and community engagement. As universities increasingly recognize the importance of food security and nutrition, future research should focus on understanding the complex dynamics of campus food systems, including sourcing, production, and waste management. Investigating the impacts of food policies on student wellbeing and academic performance can provide insights into effective interventions. Additionally, exploring partnerships with local farmers and community organizations can enhance access to fresh, sustainable foods while fostering experiential learning opportunities for students. Research should also address the implications of climate change on food sourcing and resilience in campus dining services. By prioritizing these areas, higher education institutions can contribute to more sustainable and equitable food systems, benefiting both students and communities.

Keywords Cultural sensitivity • Agriculture • Common good • Global security • Social groups • Mobility

Understanding that the college experience brings a unique diversity that has not been seen in college environments before. In the same respect food and nutrition security have both macro socioeconomic determinants,

S. Sharififard, *Community, Food Insecurity, and a Global Perspective on Campus Food Pantries*,
https://doi.org/10.1007/978-3-031-78223-7_10

such as environmental and geopolitical factors, and regional/local determinants that refer to the community environment. There are also household determinants related to education, employment, health, and mobility. Each level of determinants can affect the next generation and the makeup of the future college and university campuses. Further, how students access food, contribute to their food system, and have conversations about cultural sensitivity and food insecurity can improve communication and the quality of the college experience overall.

At the regional/local level of determination, aspects of the neighborhood where students live while in school and where they migrate after graduating significantly contributes to the social environment and the relationships that exist between individuals and groups that live and work in a community. This environment includes social cohesion, inclusivity, diversity, and belongingness. The connection and relatability between a campus, its community, and the student demographic is a critical aspect of the geography and economy of a region. Strong cohesion between the food retail environment, the college campus, and the surrounding organizations can lead to high levels of economic growth and positive change. Increasing these relationships can impact collaboration, innovation, and public health outcomes that are proactive.

Future research has the potential to explore the relationships between the types of students' academic schedules and studying patterns overall as related to consistent forms of experiencing food insecurity. Researchers can work to understand the perceptions of food access introduced in this study and consider other solutions for increasing access to healthful and sustainable foods among some of the populations experiencing food insecurity. As colleges and universities begin to offer more opportunities for students, the need for physical, social, and emotional wellness will increase. The types of foods and dietary consumption students have throughout their academic careers will also continue to develop. Such factors can affect students culturally, emotionally, and socially, as well as later in life.

Additionally, students who work while enrolled in school could be researched in focus group settings. Including focus groups in food security studies and higher education research would allow students to address their food insecurity concerns with a group of other individuals in a safe setting—the latter who have similar work and academic experiences on campus. The relationships between a student's confidence in obtaining long-term employment upon graduation and the time to complete a degree can further help with understanding academic aspirations and student success on campus.

It would also be beneficial for researchers to address systemic inequalities and ethnic preferences of food consumption. If there are adequate resources for students following a diet consistent with religious, cultural, athletic, or other personal/dietary preferences, it would be important to understand how campus food pantries supplement shortfalls, if any, of food supply for all the students they serve. Other researchers could consider identifying concentrations of food deserts and the types of food retailers commonly found in those same areas. Dietary food laws and food security policies in the United States could also be used to infer quantitative and qualitative understandings of the local food environment.

Direct changes of the retail food environment in neighborhoods and the convergence of public health concerns can also assist educational policymakers and researchers in identifying immediate risks associated with academic disparities and the identification of areas which can become food deserts. By studying dietary quality changes throughout college enrollment, policymakers, researchers, and academicians can obtain an overview of current institutional goals to align with the convergence of food systems. Additionally, campus meal plans and the types of food products available in specific regions or the neighborhoods of similarly situated populations can help inform educational policy and public health outcomes related to higher education.

Relationships among students, faculty, staff, and administration can help inform institutional policy decisions at the state and local level while advancing future development in public health and educational leadership studies. Emerging basic needs for adults and students who are returning to school could also be used to help with providing the relevant and appropriate resources necessary to succeed. The overall findings can benefit both academics and professionals who are interested in senior-level administrative positions at institutes of higher education. Further, campus meal plans and the types of food products available in specific regions or neighborhoods of similarly situated populations can help inform educational policy and public health outcomes related to higher education and equitable food systems that represent a student population.

Community food banks could help develop culturally relevant partnerships with their communities which is strongly associated with the intention to organize better food systems that align with the communities where individuals reside, work, and study. Beyond the campus, the immediate or direct changes in the retail food environment, the supply chain, and the global food system can also assist policymakers, educators, and researchers in exploring immediate risks associated with social disparities and help identify areas that can become food deserts. It is further essential to consider the best needs and outcomes for patrons. How individuals with specific nutritional needs or dietary restrictions continue to alter or improve some aspects of their regular diets without risking the quality of the foods they consume could help public health outcomes and educational and agricultural policies.

This study helped contribute to the existing research on food insecurity and campus food pantries. The researcher also focused on pantry operations and the types of barriers that students face when accessing their food needs. The researcher also looked closely at the neighborhood food environment to identify if cultural or dietary foods affect food consumption in specific areas. The study results can help with interpreting the distance students, families, and food pantry representatives travel to access or purchase foods that are necessary to operate their food pantry or consume individually. In addition to education, food and nutrition, and law, the study also helps with the emerging literature in the academic fields of public health, urban geography, and dietetics.

Additionally, the study can help with introducing new ways of approaching food insecurity and the changing food environments across communities. The study can further help students identify their consistent mode(s) of transportation and how they select their food expenditures and necessities and basic needs. While the researcher did not assess shopping behaviors and personal accessibility to specific food stores, this study's findings could help understand the relationship between the ethnic composition of specific neighborhoods and the overall diverse dietary and food preferences of student populations in the U.S.

Campus food pantries will continue to cultivate broader conversations about food security, poverty, global health, and hunger. The quality of food at campus food pantries and throughout the overall campus food environment will continue to affect students who encounter barriers to food access. The campus food pantry has become an increasingly essential function of colleges and universities. The use of campus food pantries and food banks can help policymakers identify changes in food environments and aid in developing more prosperous communities. For individual communities, it is important to be cognizant of the types of examples set forth by local academic institutions, and public entities, broadly defined. How public institutions work to support and sustain cultural, faith-based, and secular values in an environment that promotes respectful discourse, dialog, and decision-making is important to the advancement of higher education.

The university's function is to advance knowledge and provide a quality education for all students. This is possible through the focus on student success and wellbeing, cultivating a sense of community and belonging. How students secure their basic needs at a community and institutional level requires repeated storytelling, cultural awareness, empowerment, and courage. Students who are empowered to pursue the keys to their academic success can learn to thrive at their place of education with support and encouragement. It is also a collective and conscious decision to desire to make a sustainable attempt to continue having these conversations on campuses and in the communities. Universities are anchored as part of a wider community that benefits the economic and cultural capital of their futures.

Glossary

Food bank a place where food is given to people who do not have enough money to purchase them at a traditional grocer or food retailer.

Food desert a census tract that meets both low-income and low-access criteria including: (1) poverty rate is greater than or equal to 20% OR median family income does not exceed 80% statewide (rural/urban) or metro-area (urban) median family income; (2) at least 500 people or 33% of the population located more than 1 mile (urban) or 10 miles (rural) from the nearest supermarket or large grocery store

Food insecurity reports of reduced quality, variety, or desirability of diet, or multiple indications of disrupted eating patterns and reduced food intake

Food pantry a distribution center in which food, beverages, and utensils are kept and supplied for individuals in need

Food security little or no reported indications of food-access problems or limitations, or of changes in diets or food intake

Food system the inputs, activities, people, outputs, and outcomes associated with getting food from seed to plate

Nutrition security the need to secure access to an appropriately nutritious diet, comprising all essential nutrients and water, coupled with a sanitary environment and adequate health services and care to ensure a healthy and active life

S. Sharififard, *Community, Food Insecurity, and a Global Perspective on Campus Food Pantries*,
https://doi.org/10.1007/978-3-031-78223-7

Index

S. Sharififard, *Community, Food Insecurity, and a Global Perspective on Campus Food Pantries*,
https://doi.org/10.1007/978-3-031-78223-7

The manufacturer's authorised representative in the EU is Springer Nature Customer Service Centre GmbH, Europaplatz 3, 69115 Heidelberg, Germany. If you have any concerns regarding our products, please contact ProductSafety@springernature.com

Printed and bound by CPI Group (UK) Ltd, Croydon, CR0 4YY
30/09/2026
02225609-0003